URBAN WORLD/
GLOBAL CITY

David Clark

London and New York

First published 1996
by Routledge
11 New Fetter Lane, London EC4P 4EE

Simultaneously published in the USA and Canada
by Routledge
29 West 35th Street, New York, NY 10001

Routledge is an International Thomson Publishing company

© 1996 David Clark

Typeset in Garamond by
Florencetype Limited, Stoodleigh, Devon

Printed and bound in Great Britain by
TJ Press (Padstow) Ltd, Cornwall

British Library Cataloguing in Publication Data
A catalogue record for this book is available from the British Library

Library of Congress Cataloguing in Publication Data
Clark, David, 1946–
Urban world, global city / David Clark.
p. cm.
Includes bibliographical references and index.
1. Cities and towns. 2. Urbanization. 3. Sociology, Urban.
I. Title.
HV151.C588   1996   95–26472
307.76–dc20   CIP

ISBN 0–415–14436–1 (hbk)
ISBN 0–415–14437–X (pbk)

6 . 6 . £10.99

DOM
(Cla)

GLOBAL CITY

The last decade of the twentieth century marks a symbolic watershed in the history of human settlement for it encompasses a period
in which the location of the world's people has become more urban
than rural. Over half of the world's 5.2 billion people now live in
towns and cities. No longer are such centres exceptional settlement
forms in predominantly rural societies. The world is an urban place.

Urban World/Global City examines this remarkable geographical and
demographic phenomenon. It analyses the distribution and growth
of towns and cities and explores the social and behavioural characteristics of urban living. Individual chapters focus upon populations
and places, urban growth and urbanisation, urban development as
a global phenomenon, lifestyles in the city, global urban society,
world cities, and the urban future. Emphasis throughout is placed
upon the world scale, urban developments being linked to the emergence of a global economy and society. Attention is directed equally
to urban patterns and processes in developing and developed areas
which are seen as common consequences of the emergence of capitalism as the dominant economic system.

Global perspectives are increasingly important in the social sciences,
but towns and cities tend to be overshadowed by the focus upon
transnational corporations and trade. This timely overview and
analysis will be essential reading for students of geography, sociology,
planning and development studies who seek to further an understanding of the ways in which the urban world has evolved, and
how it is likely to change over the next twenty-five years.

David Clark is Head of Geography at Coventry University.

041514437X

CONTENTS

TABLES

FIGURES

ix

Note: The political boundaries which are shown on all the world maps are those which existed on 1 January 1989.

PREFACE

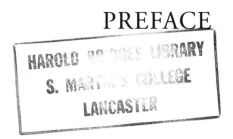

One hundred years ago, in 1899, Adna Ferrin Weber published a thesis on *The Growth of Cities in the Nineteenth Century*. This masterpiece of data collection, analysis and explanation was the first comprehensive attempt to document and understand urban development at the global scale and was responsible for establishing urban study as a central focus of inquiry in the newly emerging discipline of geography. Although cities were commonplace in that part of North America in which Weber lived and worked, they were comparatively small and thinly scattered, and together housed only a minority of the population, a pattern that was repeated across much of north western Europe. As places of residence they were different and their populations engaged in social and economic activities which were distinctively urban in character. Elsewhere in North America and the rest of the world, urban development was the exception and the population was overwhelmingly rural, in terms of both location and ways of life. It is only in the last thirty years that it has become valid to talk about an urban world: a world in which urban places and urban living are the norm rather than the exception.

The urban world is a product of three main developments. The first is that the world itself has become a coherent and integrated whole, through the globalisation of economic and social activity. Markets which were previously separate and localised have become merged, and have been superseded by world-wide patterns of production and consumption coordinated by global institutions and organisations. Social patterns and relationships which were only local in scale have become global in extent. The second development has been the growth in size, proliferation in number and spread of towns and cities into new territories, so that there are now few regions which lack urban populations and places. The third is the transformation of global society because so many people and such a large proportion of the world's population live in towns and cities and follow lifestyles which are urban in origin and character. The residual pockets of ruralism are in rapid decline. These three developments are powerful and self-reinforcing. Together they have created an urban world which would be unrecognisable to analysts writing one hundred years ago.

The contemporary urban world is, however, far from uniform. Urban growth, urbanisation and the spread of urbanism are changing the location and nature of society, but at different rates in different places. There are wide variations in the number and proportion of the population which lives in urban settlements, in the size and role of cities, and in the extent to which people live urban ways of life. The present pattern is best seen as a social and spatial spectrum with urban predominating at one extreme, rural at the other and a number of gradations in between. It is the remit of geographers to analyse and account for these contrasts and variations and to examine their consequences.

This book identifies and seeks to account for the characteristics of the contemporary urban world and global urban society. It traces the growth of towns and cities and the spread of urbanism, and attempts to explain the organisational characteristics of the global urban system. Emphasis throughout is placed upon the world scale, with variations in levels of urban development within nation states being the focus of secondary attention. What appears between these covers is an attempt broadly to overview and synthesise work on global urban patterns, trends and processes so as to trace the evolution of the urban world, its present characteristics and the ways in which it is likely to change. Little of the material is new – indeed some of it is very old – so any distinctiveness is the product of selection or organisation. Thanks are due to members of staff in the cartography unit of the School of National and Environmental Sciences, Coventry University who prepared the maps and diagrams around which the text is written, and to the undergraduate students who unknowingly acted as sounding boards for some of the arguments and ideas. Any errors of omission, over-simplification or misinterpretation are mine.

David Clark
Ashow, October 1995

1

GLOBAL PATTERNS AND PERSPECTIVES

The last decade of the twentieth century marks a major watershed in the evolution of human settlement, for it encompasses the period during which the location of the world's people has become more urban than rural. Variations among countries in the quality of their census data and in the ways in which urban areas are defined mean that it is not possible to be exact, but it is likely that 1996 was the year in which the figure of 50 per cent urban was achieved. Despite its geographical significance, this historical event has passed largely unrecognised and unreported, its profound symbolic importance overlooked. More of the 5.2 billion inhabitants of the globe now live in towns and cities than in villages and hamlets. No longer are towns and cities exceptional settlement forms in predominantly rural societies – the world has become an urban place.

Urban development on this scale is a remarkable geographical phenomenon. Far from being spread widely and thinly across the surface of the habitable earth, a population which is urban is one in which vast numbers of people are clustered together in very small areas. Whether through choice or compulsion they live in close horizontal and vertical proximity and at very high densities. They seemingly prefer, or are forced to accept, concentration rather than dispersal. The benefits of access to services and economies of scale which are a consequence of closeness and agglomeration apparently outweigh the disadvantages and drawbacks of crowding, congestion, noise and pollution. If the size of population is any guide, then living in an artificial environment seems to have greater appeal than residing in the countryside. The number and size of cities and the rate at which many of them are growing suggests that they are highly attractive and acceptable forms of settlement to most people.

This extreme form of population distribution is a product of deep-seated and persistent processes which enable and encourage people to amass in geographical space. Such mechanisms generate the surplus products upon which cities rely, and bind the urban population together by a range of economic and social ties. As such they sustain large numbers of people as viable, stable and successful urban communities. So powerful and pervasive are the forces of urban formation and growth that they presently concentrate over 2.6 billion of the world's population in towns and cities. It is difficult adequately to convey an impression of the degree of clustering which this represents, as data on the combined area of all the world's urban places are not readily available. If, however, all the urban population lived at a density of 7,600 per sq. km, which is the average for the major cities of East Asia, then they could be accommodated on less than one per cent of the world's landmass, an area roughly equivalent in size to Germany.

The urban world is equally distinctive in socio-economic terms. Despite the infinite and intricate variations of tradition and culture that exist within and between nations, cities appear to have and to be acquiring more in common than they have differences. Urban places have many similarities of physical appearance, economic structure and social organisation and are beset by the same problems of employment, housing, transport and environmental quality. The elements in many urban skylines are the same, as commercial and residential areas are increasingly dominated by high-rise developments constructed in international styles. Streetscapes across the world are adjusting in the same way to accommodate the needs of the ubiquitous car, so that cities are fast losing their individual layouts and architectural identities. Within buildings, workers do the same sorts of jobs, often on the same makes of computer or machine, and manufacture goods and services to the requirements of world markets dominated by a small number of global producers. Patterns of demand are converging as consumerism absorbs ever more of the world's population. There are few cities where McDonald's hamburgers, Fuji films, Levi jeans and Coca-Cola are not readily available and purchased in quantity. Some of these similarities are superficial and hide important cultural differences, but the underlying trend is clear. Visual, anecdotal and research evidence indicates that there is increasing convergence among cities in economic and physical terms.

Irrespective of continent or country, many urban residents live their lives in broadly similar ways, with common concerns over home, children, school and work. Attitudes and expectations are shared as many aspire towards the lifestyles which are popularised and promoted by the mass media. Billions of people feast nightly on a diet of televised soap operas and international sporting events, with pop singers, film stars, sports personalities and media celebrities enjoying a world-wide recognition and following. Such interests, fads and tastes are increasingly independent of ethnicity, colour, class and creed. They draw together and fuse what geography and culture traditionally separate and divide. The contemporary urban world is more than a motley assemblage of diverse settlements. Many observers argue that it is slowly becoming a unitary and uniform place, a global city in which most of its inhabitants are imbued with a similar set of all-encompassing urban attitudes and values, and follow common modes of behaviour.

Although towns and cities have existed for over eight millennia, the wholesale transition to urban location and urban living is very recent in origin. Many highly successful urban civilisations existed in the past, but their impacts were both limited and localised. In 1700 fewer than 2 per cent of the world's population lived in urban places and these were concentrated in a small number of city-states. Major and rapid changes began in Britain in the late eighteenth century in response to the locational dictates of industrial capitalism. They subsequently spread to north western Europe and north eastern USA, so that by the beginning of the twentieth century about 15 per cent of the world's population was living in urban places. Urbanisation as a global phenomenon is, however, essentially a feature of the last half of the twentieth century, indeed of its last three decades. Large parts of the world were effectively untouched by urban development and urban influences until 1970. It is occurring today because of massive changes in the distribution of population in countries which until recently were substantially and profoundly rural. Contemporary processes of urban development affect vast numbers of people across the globe. Although it is taking place at a local level, the present switch from rural to urban constitutes the largest shift in the location of population ever recorded.

Historical milestones are occasions for reflection and speculation; for looking back and to the future. They are a time for assessing what has been achieved and what opportunities and obstacles lie ahead. The emergence of an urban world and the prospects for a global city pose important questions concerning the nature and consequences of the urban pattern and experience. They focus attention upon the reasons how and why cities exist, the ways in which they grow, and their impact upon society. They raise issues concerning spatial and temporal variations in levels and rates of urban development and the implications of future urban change. Such concerns challenge social scientists to develop appropriate philosophies and methodologies by means of which the urban world can be conceptualised, explained and understood. The fact that the proportion of humankind that lives in towns and cities is rising rapidly further focuses attention on the likely nature of the urban world in the early decades of the twenty-first century and the characteristics and consequences of global urban society.

STUDYING THE URBAN WORLD

In view of this ambitious research agenda, it is not surprising to find that the task of analysis and explanation occupies an army of specialists drawn from a wide range of fields in the social and environmental sciences. No single discipline can claim to monopolise the study of the city, since urban questions and problems cut across many of the traditional divisions of academic inquiry. Equally, no single methodology predominates in urban analysis, for the complexities of urban life necessitate the adoption of a wide variety of approaches. It is in the interdisciplinary nature of urban issues that the city poses the greatest difficulties to the analyst. Progress in urban study requires the fusion of insights derived from a number of subject areas, each of which approaches the analysis of urban settlements in its own distinctive way.

Geographers are prominent among the researchers who set out to analyse and explain the urban world and the global city. In focusing upon location they seek to add a spatial dimension to the understanding and interpretation of global urban phenomena. Geographers are concerned to identify and account for the distribution and growth of towns and cities and the spatial similarities and contrasts that exist within and between them. They focus upon both the contemporary urban pattern and the ways in which the distribution and internal arrangement of settlements have changed over time. Emphasis in urban geography is directed towards the understanding of those social and economic processes that determine the existence, evolution and functional organisation of urban places and the characteristics of urban society. In this way, geographical analysis both supplements and complements the insights provided by allied disciplines in the social sciences which recognise the urban world as a distinctive focus of study.

A wide range of approaches is used by geographers in urban study. Simple mapping of distributions is the starting point for most geographical work; it identifies basic patterns and draws attention to possible causal relationships. It is especially appropriate at the global scale where it is important to start with an overall picture, and where variations, and hence the implications of urban development, are both complex and pronounced. Mapping, however, leads to little more than low-level description and is undermined by the highly variable quality of global urban data (see Appendix). Explanation is facilitated if attention is concentrated upon causal actions and mechanisms. Such relationships are products of tradition, culture and political organisation and are deeply embedded in the underlying strata which give societies and economies their form. A key task for geographers is to investigate and understand the structural relationships which give rise to processes which in turn are responsible for creating observed urban patterns.

Although an urban specialism is long established in geography, the adoption of a world perspective on cities and society is a recent development. It was foreshadowed nearly a hundred years ago in the formative work of Weber (1899) on *The Growth of Cities in the Nineteenth Century*, although the urban 'world' which he analysed consisted of only around fifty countries in which there was significant urban development. This major empirical study was important because it showed how much earlier and further advanced were England, Wales and Scotland as urban societies and how little urban development there was outside north west Europe and the eastern seaboard of the United States of America. The world at the time was very much a rural place, in which the development and distribution of towns and cities was limited and urban influences were restricted and localised. Within urbanised countries, however, statistics on employment patterns, family structures and demography pointed to the existence of pronounced urban–rural contrasts. Cities were places with particular socio-economic characters which sustained and perpetuated distinctive patterns of social and economic behaviour. They were places in which urban ways of life evolved and were spread into the surrounding rural areas. As well as identifying the salient characteristics of the urban world, Weber's analysis was of considerable significance in a technical sense, since it drew attention to the many problems of data availability, quality and comparability that so bedevil urban analysis and understanding at the global scale.

The principal focus of subsequent work in urban geography was on systematic themes rather than overall global patterns. Important contributions were made to the understanding of urban lifestyles (Wirth, 1938), city size distributions (Zipf, 1949; Berry, 1961) urbanisation (Davis, 1965) and the colonial city (McGee, 1967), establishing traditions of research in these areas, but these studies directed attention towards constituent parts and individual phenomena rather than the urban world as a whole. Nowhere is this more evident than in the wealth of literature on urban growth and urbanisation as recently overviewed and synthesised by Brunn and Williams (1993). This approach invariably concentrates upon either the developing or the developed world; or else adopts a country-by-country or region-by-region approach. Examples of the former include texts on urbanisation in the Third World by Breeze (1966), McGee (1971), Abu-Lughod and Hay (1979), Potter (1985), Gugler (1988) and Gilbert and Gugler (1992). Country studies are exemplified by studies of urbanisation in Africa (O'Connor, 1983), India (Ramachandaran, 1993), and China (Kirkby, 1985). Such work has generated some powerful insights into the underlying patterns and processes, but the regional and national focus diverts attention away from wider similarities, possible common causation, and the relevance of general theory.

GLOBAL PERSPECTIVES ON ECONOMY AND SOCIETY

The global view involves a focus upon scale and process, since it is concerned with world-wide patterns and the mechanisms which create them. There is nothing new in looking at the world as a whole, but it is only recently that analysts have suggested that social and economic relationships at this level can be attributed to a small number of powerful forces which operate globally. This approach derives much of its contemporary popularity from the work of Immanuel Wallerstein on the nature and extent of the links between societies today and in the past. For Wallerstein (1979, 1980, 1989), most of the discrete localised economies which once existed are now merged and amalgamated into a single, integrated, world-economy. This is global in organisation and reach and incorporates and encompasses the majority of the world's nation states and territories. Few regions lie outside its limits and are untouched by its influence. The world economy is capitalist in formation, in that it is based upon principles of private rather than state ownership of the means of production and seeks to generate profits through the manipulation of land, labour, finance and entrepreneurship. It is primarily concerned with making and providing goods and services, be they computer equipment, cars, televisions, food, textiles, weapons or media products, for global consumption. Patterns of supply and demand, circulation and exchange, and marketing and advertising, together with all the economic and social structures which make them possible, are world-wide rather than local in scale. Social and economic outcomes in the form of opportunity, advantage, injustice and poverty are similarly global in extent and implication.

The world economy is controlled by a small number of powerful transnational corporations which dominate global networks of production and consumption. They have a disproportionate influence over supplies of raw materials and manufacturing capacity, and determine and direct patterns of spending through advertising and promotional activities. They dominate and control the key economic sectors. Transnational corporations are supported by banking and investment institutions that manage and manipulate global finance, and by a range of organisations that provides producer services in the form of management consultancy, legal, personnel and marketing advice, on an international basis. The organisation of the world economy is made possible by and is maintained through an international division of labour in which the tasks which people perform, their working conditions and their rates of pay are determined by the requirements of global capitalism.

Although it functions as an integrated whole, there are important structural variations within the world economy. In geographical terms, the world-economy consists of a set of dominant core states in which most innovation and advanced activity takes place, and a dependent periphery and less dependent semi-periphery area characterised by low-level production, low wages and coerced labour (Figure 1). Countries in the core have relatively high incomes, advanced technology and diversified production. They are generally prosperous and their populations enjoy high standards of living and qualities of life (Knox and Agnew, 1994). Those outside the core are less well developed and have economies which depend upon primitive technology and undiversified production. Many, especially in the periphery, are poor and are severely deficient in infrastructure

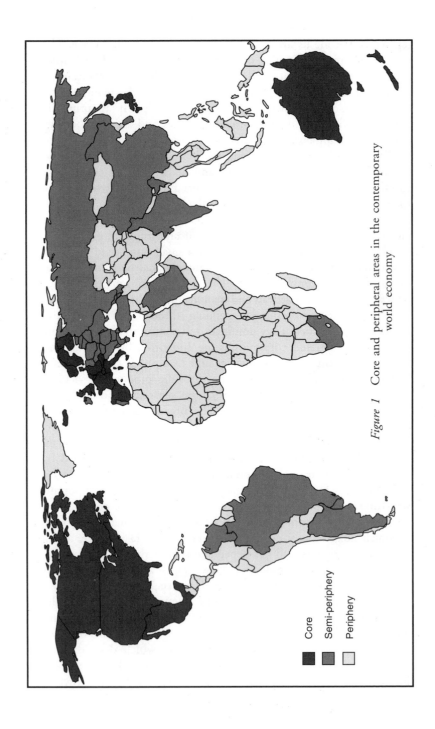

Figure 1 Core and peripheral areas in the contemporary world economy

Core

Semi-periphery

Periphery

and social provision. World systems theorists argue that the core achieves its greater prosperity through economic and political domination and control over peripheral and semi-peripheral areas. This relationship, as with the geographical consequences which follow from it, is maintained and perpetuated over long periods of time through the evolution and changing organisation and spatial relationships of capitalism.

The world-economy is organised around and through cities. Implicit in the global approach is the view that cities must increasingly be seen as interacting and interrelated elements within an urban hierarchy that underpins and makes possible processes of capitalist accumulation and reproduction (Timberlake, 1987; King, 1990). Rather than merely act as points of exchange for goods and services produced and consumed in their surrounding areas, which was the historical pattern, they are places of articulation where people and products link to the wider world. Cities are locations through which global goods and services reach their markets and are consumed, and from which surplus values are extracted. They interlock and intermesh in the form of local and national networks which in turn are incorporated within a global urban hierarchy. This system is dominated by a small number of world cities, housing the headquarters of the principal transnational corporations, finance and producer service organisations. Such centres are effectively the command and control points for global capitalism.

The emergence and spatial organisation of the urban world, according to this perspective, are dictated by the needs of the world economy. The accumulation of wealth through manufacturing, exchange and consumption is the primary cause of urban growth and urbanisation. It leads to a concentration of population in towns and cities throughout the core and in the periphery so that urban development in both is an interdependent outcome of the operation of global capitalism. Spatial and temporal variations in levels of urban development are consequences of the ways in which capitalism has evolved and its changing relations with areas of supply and demand across the world. Overseas resources and markets are secured and manipulated by colonisation, colonialism and imperialism. Explanations of urban development lie in the social and economic characteristics of successive forms of capitalism.

As well as increased economic integration and interdependency, the world has moved closer together in social terms. Social networks which once were closed and localised have become open and interconnected, fusing at broader spatial scales. Globalism has replaced parochialism. Society has become wider, in the sense that many people across the world now share and subscribe to the same set of values and beliefs and have a broadly similar social experience. This development is made possible by advances in transport and communications which overcome barriers of inaccessibility and distance and facilitate easy and cheap world-wide movements of people and ideas. It is reflected and reinforced by the international traffic in tourists, business travellers, migrants and workers, and in the long-distance and instantaneous circulation of information and imagery by broadcasting, telecommunications, videos and the world-wide web of interconnected computers. Such exchanges and flows bind individuals into interest groups and communities which extend across national and regional boundaries. They facilitate and encourage the creation of social systems which are not proscribed by geography. Society has become divorced from space. Membership and role are functions of participation and not place.

A case can be made that global society is increasingly urban in character. Cities are points of production and reproduction of urban culture. As major and dense concentrations of population drawn from many different backgrounds, they are places in which a diverse array of beliefs, styles, values and attitudes originate, ferment and flourish. These combine in the form of patterns of association and lifestyle which are distinctively urban in character and differ markedly from those which exist in rural areas. Such modes of thought and behaviour are carried and spread by movements of people and flows of information and ideas well beyond city boundaries so that they influence and can be adopted by populations across the world. Human society is becoming urbanised in the sense that increasing numbers of people are being exposed to and are absorbing the social values which arise out of and are most closely associated with life in cities. Some of the key questions in urban geography concern the nature of global urbanism and the ways and extent to which it interacts with and so may be destroying, modifying or reinforcing traditional local cultures.

CONCLUSION

This book examines the cities of the world and the world as a city. It seeks to identify and account for the growth and organisation of the contemporary urban world and the causes, characteristics and consequences of global urban society. Emphasis throughout is placed upon the global patterns which appear when data are compared and mapped nation by nation, and upon the world-wide processes which, many analysts argue, are responsible for them. Variations in levels of urban development and urban character within nation states, which of course may be pronounced in large countries, are a focus of secondary attention. The approach is largely synoptic, the purpose being to draw together, overview and synthesise the literature, both established and recent, so as to show how an understanding of the urban world can be approached and has evolved.

Given the highly ambitious goals of analysing and explaining the location and behaviour of half of the world's population, the treatment is necessarily and deliberately broad. The aim is to provide an introduction to the urban world and the global city by illustrating and exploring the underlying processes, relationships and issues that constitute the subject of more detailed and advanced study. Many questions are raised but few answers are provided, reflecting the innate difficulties of generating explanations and understanding at the world scale and the variable quality and availability of published research on global urban issues.

Individual chapters advance the overall aims by focusing in turn upon historical patterns of urban growth and urbanisation (Chapter 3), urban development as a global phenomenon (Chapter 4), place-bound interpretations of lifestyles in the city (Chapter 5), global urban society (Chapter 6), world cities (Chapter 7), and the future urban world (Chapter 8). The definitions and data which are so critical in urban analysis and upon which global urban study is based are discussed in detail in the Appendix. Having outlined the aims and objectives in general terms, Chapter 2 now provides an overall context by using simple census statistics and elementary theory to outline and explain the contemporary distribution of urban populations and urban places at the global scale.

RECOMMENDED READING

Bassett, K. and Short, J. (1992) 'Development and diversity in urban geography', in Gregory, D. and Walford, R. (eds) *Horizons in Human Geography*, London: Macmillan, 175–93.

A comprehensive and up-to-date review of the field of urban geography which identifies and explains different theoretical approaches and evaluates their advantages and limitations.

Herbert, D. T. and Thomas, C. J. (1990) *Cities in Space, Cities as Place*, London: Wiley.

Provides a comprehensive introduction to the major topics of inquiry in urban geography, with particular strengths in the social characteristics and problems of cities.

Janelle, D. G. (1991) 'Global interdependence and its consequences', in Brunn, S. D. and Leinbach, T. R. (eds) *Collapsing Space and Time: Geographic Aspects of Communication and Information*, London: Harper Collins, 49–81.

Provides a useful introduction to developments in global communication and information flow and the implications for the emergence of global urban society.

Johnston, R. J., Taylor, P. J. and Watts, M. J. (eds) (1995) *Geographies of Global Change*, Oxford: Blackwell.

An invited collection of papers by leading geographers on a wide range of aspects of global distributions and global change in the late twentieth century. The editors provide useful contextual introductions and conclusions.

Knox, P. and Agnew, J. (1994) *The Geography of the World Economy*, London: Arnold.

An excellent introduction to the concept of the world economy and the ways in which an understanding of contemporary economic development can be advanced through the adoption of a global perspective.

Williams, J. F. and Brunn, S. D. (1993) 'World urban development', in Brunn, S. D. and Williams, J. F. (eds) *Cities of the World*, London: Harper Collins, 1–41.

A useful introduction to the key concepts and definitions in urban study and to the main structural features of the contemporary urban world.

2

URBAN POPULATIONS
AND PLACES

The urban world is a heterogeneous place. Although in geographical terms the population of the globe is more urban than rural, levels of contemporary urban development vary widely. Important and highly significant differences exist within and between regions and countries in the size and proportions of their populations that live in urban places. The ways in which the population is distributed according to the number and size of cities also differ markedly from place to place. Such differences are the products of the many complex processes that are responsible for the contemporary pattern of global urban development and change. They have important implications for the ways in which the world integrates and functions as an urban system. Existing geographical differences combined with varying rates of growth constitute powerful pointers to the ways in which the urban world is likely to evolve over the next thirty years.

Generalisations concerning the urban world depend for their validity upon data which relate to each of the world's sovereign states. They are highly sensitive to definitions of urban places, the size of countries, and the timing and accuracy of national population censuses. Particular problems surround the reliability and frequency of censuses in many of the world's poorest nations where, paradoxically, the scale and rates of contemporary urban change appear to be greatest. They are especially acute in highly populated countries, where even small errors of specification and enumeration may result in the misclassification or omission of many millions of people. The quality of censuses throughout much of Africa is especially poor, so that the numbers and rates of growth of the urban population are in many cases matters of informed speculation. International urban statistics are surrounded by many difficulties of availability

13

and reliability and must be regarded as crude estimates rather than precise measures. Rather than refer to data problems continually throughout the text, they are examined in detail in Appendix 1.

Some indication of the difficulties that can arise is shown by the statistics on urban China. The definition of 'urban' as followed in the State Statistical Bureau's Statistical Yearbook is based upon a range of criteria which includes place of residence, length of residence, place of population registration, the status of suburban counties and an individual's source of grain supply (Goldstein, 1989). On this basis some 440 million people, amounting to 33 per cent of the Chinese population, were classified as urban in 1990 (United Nations, 1991). A characteristic of Chinese cities, however, is that they frequently annex a large number of adjacent districts into their administrative areas in order to ensure control of vital urban supply needs, such as reservoirs or power plants. As a result, many Chinese cities incorporate large rural areas, the populations of which live many miles from the city yet inflate its official population. Following this wider definition, the World Bank suggests that in 1989, some 560 millions or 58 per cent of the population of China were urban (World Bank, 1992). The difference of some 120 million between the State Statistical Bureau and World Bank figures is greater than the total urban population of Sub-Saharan Africa.

Although it is important to be aware of the major shortcomings, there is little that can be done to compensate for deficiencies and to allow for differences. The most sensible course of action, and the one followed in this book, is to seek consistency by drawing upon a single data set with known characteristics. Detailed information on the urban world in the form of population estimates is assembled and published on a regular basis by the United Nations in its annual *Demographic Yearbook* (United Nations, 1994) and in its *World Urbanisation Prospects* (United Nations, 1991). Both incorporate data for each of the world's sovereign states which are principally based upon national censuses. It is by reference to these data, and especially the latter, in which urban populations are estimated for 1995, that the key features of the urban world, and the issues that surround them, are most easily and reliably identified and explored.

The geography of the contemporary urban world is characterised by pronounced variations in the number and proportion of people who live in urban places. Some parts of the world have huge numbers of urban people; in others there are very few. The map of urban

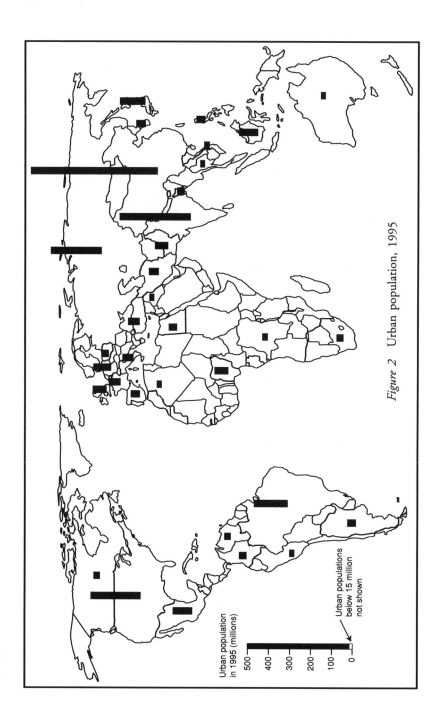

Urban population
in 1995 (millions)

500
400
300
200
100
0

Urban populations
below 15 million
not shown

Figure 2   Urban population, 1995

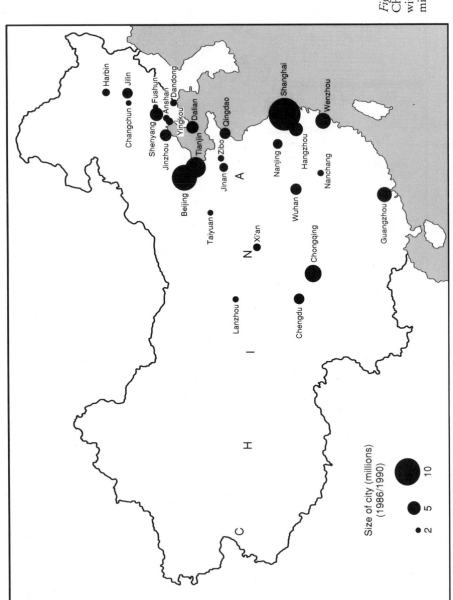

Figure 3
Chinese cities
with over two
million people

population like the map of total population is dominated by China and India (Figure 2). China has by far the largest number of urban dwellers and as such merits particular attention in any balanced analysis of the contemporary urban world. One in five of the world's urban people lives in China and the total population of Chinese towns and cities, at 499 millions (in 1995), is similar to the urban population of Africa and South America combined. Chinese cities are both numerous and large, there being 28 with populations in excess of 2 million and 46 with between 1 and 2 million (Figure 3). The urban population of India is some 279 millions, almost exactly the same as the urban population of the whole of Africa.

The urban populations of China and India are so high that they completely overshadow and suppress the more subtle variations which exist elsewhere and which may be highly significant at the local scale (Figure 2). It is impossible to portray the urban populations of China and the Gambia on the same map! Sizeable urban populations occur in the USA, the former USSR, Mexico and Brazil because these countries are large and because high proportions of their populations are urban. Japan and Indonesia are smaller countries with large urban populations. An urban population of some 378 millions is spread across the 36 small nations which comprise the continent of Europe (excluding the former USSR). Comparatively few people, however, live in towns and cities in Africa. This is both because of sparse population overall and because the percentage of the population which lives in urban places is low.

The distribution of the world's urban population is not reflected in the balance of research activity in urban geography. Most research has been undertaken on the towns and cities of the USA and Europe, although the majority of the urban population live outside these areas. It has generated a body of theory and concepts which are based upon and so inevitably to a degree reflect Western values and experience. The urban geography of China is especially under-researched and knowledge of Chinese cities and the ways in which they grow is fragmentary. The reasons are to do with the restrictions which are placed upon social and economic study in a totalitarian state, and the paucity of published data available to researchers outside China (Sit, 1985; Goldstein, 1989). Until recently China was closed to foreign academics. None of the authors who contributed to the three seminal volumes of collected essays on Chinese cities edited by Lewis (1971), Elvin and Skinner (1974), and Skinner (1977) had been allowed to visit the country and it was not until the borders were opened in 1978 that urban field-work was possible. A complete list of annual figures for urban population compiled on a comparable basis was not available until 1983 (Sit, 1985). There is in contrast a rich and long established tradition of urban geographical study in India, based upon the excellent Census of India, which traces its origins to the work of R. L. Singh at the University of Delhi in the 1950s. Of particular importance are the contributions of Turner (1962) and Ramachandaran (1993). The volume, however, is small in comparison with the size and scale of India's cities and urban problems. Such variations in levels of research effort detract significantly from the ability to form a balanced and informed view of the contemporary urban world.

Some of the highest levels of urban development in proportionate terms are found in South America, the most urban continent (Figure 4). The population is more urban than rural in all but one of the South American countries (Guyana), and over 80 per cent of the population of Venezuela, Uruguay, Chile and Argentina are town and city dwellers. The proportion of the population that is urban is similarly high in Europe, Western Asia and Australasia. It is over 80 per cent in the United Kingdom, the Netherlands, Belgium, Germany, Denmark, Saudi Arabia, Israel, Australia and New Zealand. Belgium, with some 97 per cent of its population living in towns and cities is, with the exception of city-states such as Hong Kong and Singapore, the world's most urban country. Unlike the Americas, where levels of urban development are uniformly high, there are, however, countries in Europe and Western Asia which are predominantly rural. Examples are Portugal, Albania, Yemen and Oman.

Levels of urban development are low throughout most of Sub-Saharan Africa, South and East Asia. Only a small percentage of their populations lives in urban places and these regions include many of the world's most rural areas. The village is the most common unit of settlement and towns and cities are the exception rather than the norm. Fewer than one person in three in Sub-Saharan Africa is an urban dweller. The figure is less than one in ten in Burundi and Rwanda and less than one in five in Ethiopia, Malawi, Uganda and Burkina Faso. Countries in this region have small populations and the percentage that lives in urban places is low. The prevailing pattern in these areas is one in which the population is distributed so widely, and in very small settlements, that an articulated urban system, capable of supporting those modern facilities and services that require large concentrations, has yet to emerge (Rondinelli,

18

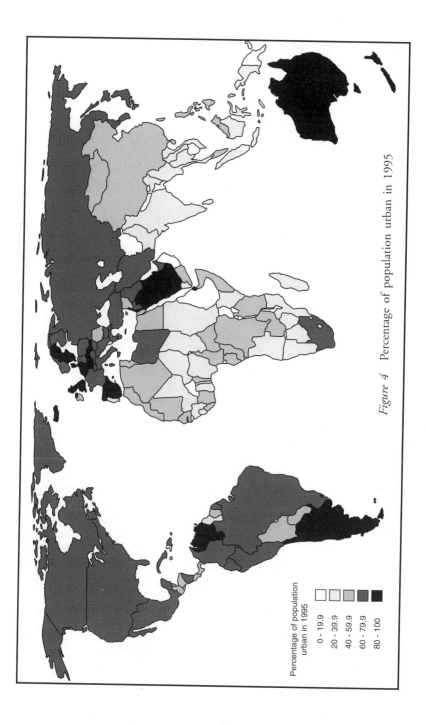

Figure 4 Percentage of population urban in 1995

Percentage of population
urban in 1995

0 - 19.9
20 - 39.9
40 - 59.9
60 - 79.9
80 - 100

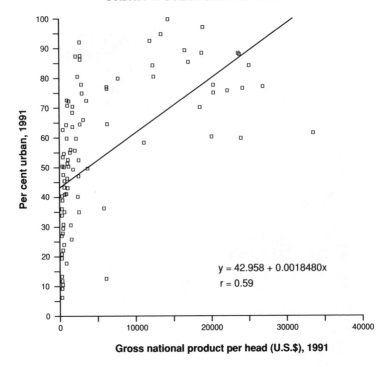

Figure 5 The relationship between urban and economic development

1989: p. 294). The situation in South and East Asia is different because it contains the world's most highly populated countries, although the distribution of population within these countries, and in the region as a whole, is predominantly rural. With some 30 per cent of India's 930 million people thought to be living in urban places, the level of urbanisation there is similar to that in China and Pakistan. Fewer than 25 per cent of the population of Cambodia, East Timor, Laos, Vietnam, Afghanistan, Bangladesh, Nepal and Sri Lanka is urban. The Himalayan kingdom of Bhutan is reckoned to be the world's most rural sovereign state, with only 6 per cent of its population living in towns and cities.

It is important to emphasise the complexity and variety of contemporary urban distributions and the fact that this undermines the value of generalisations concerning the urban geography of continents or regions. A particular complication is that they correspond only approximately with simple divisions of countries into 'developed' and 'developing'. The number and proportion of the population that live in towns and cities are products of a country's history, culture and resources and are only weakly linked to its level of contemporary economic development. The correlation between gross national product per head of population, and the percentage urban, is 0.59 (Figure 5). Countries with high gross national products per head tend to have high levels of urban development, but there is a very wide range of levels of urban development among countries with low GNPs per capita. For example, the GNP per capita in Namibia and Peru in 1994 was below US$2,000 but the proportion of the population living in towns and cities was 70 per cent and 27 per cent respectively. Conversely, Austria and Ecuador have a similar proportion of their population in urban places (54 per cent) but their GNPs per head differ by US$19,140. The relationship is distorted by the fact that urban populations and levels of urban development are high throughout South America and yet the countries in that continent occupy only middle rankings on the World Bank's indices of economic development. South America has an urban history that is very different to that of other parts of the developing world. The data show that the developing or Third World is highly differentiated in urban terms. There are as many variations of urban development within the developing world as there are differences of urban development between it and the developed world.

A WORLD OF TOWNS AND CITIES

The urban population is distributed among settlements of widely differing size. Urban places range from towns and cities with several thousands of people to those with tens of millions. An extensive vocabulary of descriptive labels is available to describe these different concentrations, although the terms lack precision and consistency of meaning from country to country. It is difficult to make clear distinctions in size terms because urban places are members of a continuum which grade into one another. Basic distinctions can be drawn between towns and metropolises, and between cities and megalopolises, but the differences are not so sharp between adjacent members of the continuum. Most would agree that settlements with populations over 100,000 are probably cities, but the status of places with around 200,000 is more questionable, especially when they have important local government and commercial functions. As one goes down the scale from the largest urban agglomeration to the smallest

town it is extremely difficult to identify break points and termin-
ology that are universally acceptable.

Most of the world's urban population live in small to medium-
sized urban places. Despite the emphasis in the literature which is
placed upon megalopolises and mega-cities, it is important to emphasise that the majority of the urban residents live in settlements whose
populations are measured in thousands rather than in millions.
According to United Nations estimates, some 66 per cent of the Third
World's urban population are in cities with fewer than one million
inhabitants. The Census of India records some 75 per cent of the
urban population in places with populations below one million. The
equivalent figure for China is 84 per cent (Goldstein, 1989: p. 208).

The primary function of most of these intermediate settlements
is to act as points of linkage between town and country where agricultural surpluses are exchanged for urban goods and services. The
importance of this role is reflected in the siting of such cities in
highly accessible positions within local transport networks, and their
level and range of government, administrative and commercial functions. Such activities, and their spatial arrangements, are explained
by central place theory, the general validity of which is verified by
numerous studies of market centres and retail distribution networks
and relationships across the world (Clark, 1982). Some of the most
important research questions in urban geography surround the ways
in which these intermediate and market centres are integrated within
national and global urban systems and the consequences of such
involvement for the lifestyles of their residents.

A notable feature of the contemporary urban pattern, however, is
the degree to which the urban population lives in giant cities (Dogan
and Kasarda, 1989). In 1990 the United Nations recognised 270
cities with over a million people, which together housed 33 per cent
of the world's urban population (United Nations, 1991). There were
22 cities with between 5 and 8 million residents. Some 10 per cent
of the urban population lived in 20 mega-cities, or urban agglomerations, of size 8 million or more (Figure 6). The best estimate is
that Mexico City had some 20 million inhabitants in 1990, while
Tokyo had 18 million, São Paulo 17.4 million, and New York some
16 million. The way in which the population is distributed among
cities of different size has important geographical implications. Not
only is the urban population concentrated in a small number of
countries but, within many of these countries, there is a disproportionate concentration in a small number of cities.

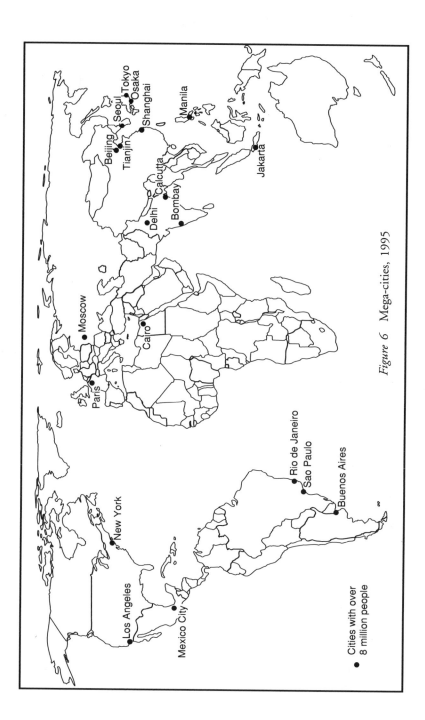

Figure 6 Mega-cities, 1995

Cities with over
8 million people

The concentration of the urban population of countries into large cities occurs in all parts of the world. Again, it is a pattern which is independent of region, length of urban history, and level of economic or urban development. Metropolitan dominance is most pronounced in Latin America and the Caribbean. The 'million-cities' in this region house 45 per cent of the urban population. They include Mexico City, São Paulo and Buenos Aires, which are believed to be the largest, the eleventh largest and the eighteenth largest urban agglomerations in the world respectively. Million cities equally dominate the urban hierarchy in many of the world's most rural regions and countries. Despite the low overall level of urban development, around 30 per cent of the urban population of South Asia live in cities with over one million people. They include Calcutta, Bombay and Delhi, cities which are among the twenty-five largest in the world. A similar percentage live in million cities in Sub-Saharan Africa. Lagos is the largest city in this region, with a population of around 6 million.

Despite their enormous size, the world's major cities at present are viable and stable places which represent a significant social and economic achievement. They contribute disproportionately to national economic growth and social transformation by providing economies of scale and proximity that allow industry and commerce to flourish. They offer locations for services and facilities that require large population thresholds and large markets to operate efficiently. The major cities house many millions of people at extremely high densities and yet provide a range of opportunities and quality of life which is greater than that which is enjoyed in the surrounding rural area. Urban residents, even in the world's poorest countries, have better access to services and higher levels of social welfare than those who live outside the city. The differences are reflected by most of the major social indicators. For example, infant mortality rates are lower in urban areas in eighteen of the twenty-two developing countries for which comparable statistics are reported by Gilbert and Gugler (1992: p. 67). The urban rate is more than twenty per thousand lower than the rural rate in Brazil, Ecuador, Ghana, Indonesia, Liberia, Mali, Morocco, Nigeria, Peru, Senegal, Thailand, Togo and Zimbabwe.

Dire warnings of the imminent social or economic collapse of one or other mega-city appear periodically in the press, but such a disaster has yet to happen. Social conflict or economic catastrophes are normally played out across and between regions or nation states, rather than exclusively within individual cities. Disturbing pictures of poverty, congestion and pollution in Calcutta, Bombay, Rio and Bangkok, and riots in Los Angeles and Beijing readily divert attention from what such places represent. Rather than gigantic social mistakes, cities, generally, are highly successful settlement forms.

Mexico City, the world's largest urban agglomeration, houses a population of some 20 million people in an area of 1200 sq. km. It is the principal economic and political centre in the country and has the greatest concentration of industry, the largest consumer market and the most highly trained workforce. Despite its size it is a stable and relatively prosperous community. Some 80 per cent of homes have a piped water supply, and 70 per cent are on mains sewerage. Over one thousand immigrants arrive in the city and are absorbed into the urban economy and society each day (Schteingrat, 1990). Unemployment, underemployment, poor housing and congestion are significant problems, but they are less severe than in rural areas of Mexico. The city's principal difficulties, paradoxically, are physical rather than socio-economic, and stem from the poor initial choice of a site, at over 2300 metres above sea level and in an area prone to earthquakes (Griffin and Ford, 1993). It is also affected by frequent temperature inversions which tend to cause a build-up of atmospheric pollution. Mexico City is far from being an urban utopia, but its existence and growth underlines the viability and attractiveness of major concentrations of population.

CITY SIZE DISTRIBUTIONS

Some of the most important questions in the study of the contemporary urban world surround the reasons why the population in a country is distributed among cities of different size. Implicit in the concept of the urban hierarchy is the assumption that population is spread across a range of cities of different sizes which interact and interdepend as a functioning urban system. Some disagreement exists among analysts concerning the normal shape of such a distribution but there is general support for the view that something approaching a log normal pattern characterises a mature urban hierarchy (Hawley, 1981). For example, it was observed half a century ago by Zipf, among others, that in many countries there is a regular gradation of cities according to size (Zipf, 1949). In fact the second city was invariably, according to Zipf, half the size of the first, the third city a third the size of the first and so on, so that the size of any centre

Table 1 Levels of urban primacy, 1991

| Country | Largest city | Population in largest city as percentage of: | |
		urban population	total population
Singapore	Singapore	100	100
Hong Kong	Hong Kong	100	94
Guinea	Conakry	89	23
Mauritania	Nouakchott	83	39
Burundi	Bujumbura	82	4
Costa Rica	San José	77	36
Congo	Brazzaville	68	28
Angola	Luanda	61	17
Thailand	Bangkok	57	13
Gabon	Libreville	57	26
Liberia	Monrovia	57	26
Haiti	Port au Prince	56	16
Greece	Athens	55	34
Togo	Lomé	55	14
Rwanda	Kigali	54	4
Jordan	Amman	53	32
Kuwait	Kuwait City	53	20
Senegal	Dakar	52	20
Laos	Vientiane	52	10
Sierra Leone	Freetown	52	17

Source: World Bank: World Development Report, 1991

could be predicted simply by its rank and the size of the largest place. So widespread was this relationship thought to be that it became known as the 'rank-size rule'. Of thirty-eight countries studied by Berry in his highly influential paper in 1961, the cities in thirteen were found to follow a rank-size pattern. The United States, the United Kingdom, Japan, China, Indonesia and Brazil are examples of countries in which the distribution of cities is presently rank-sized (Brunn and Williams, 1993).

The situation in many countries is, however, very different in that the population is unevenly distributed among urban places. The extreme is reached when there is one excessively large primate centre which dominates all the others, a pattern which was first recognised in several countries by Jefferson in 1939. Primate cities are not necessarily large in international terms but they are, by definition, very much bigger than any other place in the country. Primacy may be measured in terms of the relative size of the largest city, its share of the national population, or via a number of composite indices such as those devised by Walters (1985) which seek to take account of the deviations of observed city sizes from the sizes that would be expected on the basis of the rank-size rule. Taking national population share as the criterion, it is apparent that primate urban distributions occur in many countries, though they are most common and most pronounced among the poorest and most sparsely populated states in the developing world (Table 1). Apart from Singapore and Hong Kong, in which there is only one city where all the urban population lives, primacy is most prevalent and most extreme in Africa. Eleven out of the twenty most primate countries are in that continent. Over 80 per cent of the urban population of Guinea, Mauritania and Burundi live in the largest city. Such is the dominance of the primate city that in some countries it houses a disproportionate share of the total as well as the urban population. In the case of Guinea, Mauritania, Costa Rica, Congo, Gabon, Liberia, Greece and Jordan, this figure is in excess of 20 per cent.

Primacy has an important spatial dimension. As well as the housing of a disproportionate share of urban and national populations, it means that in many cases most of the total population is concentrated in one small part of a country. Many primate cities, moreover, are coastal ports so that the population is peripherally rather than centrally located within the boundaries of the nation state. This applies to fourteen of the countries in Table 1. Such organisational and locational arrangements have far-reaching implications for internal integration and for the ease with which social and economic developments are likely to spread from the primate city to more distant regions.

City size distributions have been extensively studied, both over space and over time. Although there is no overall consensus, research supports two broad generalisations. The first is that rank-size or log normal patterns are most common among mature, well integrated and balanced economies, while primate distributions are more normal in the developing world (El Shakhs, 1972). The second is that the degree of primacy decreases over time: as the economy matures, so the population becomes more evenly spread across all the cities (Chase-Dunn, 1985). One city may remain dominant, but not excessively. Such findings suggest that city size distributions may relate in some way to the degree to which a country's cities are integrated within the global urban network.

A useful conceptual framework, first introduced by Berry (1961), is that of the urban system consisting of cities and the links between them. Such links may be flows of people, raw materials, goods, finance, information and ideas. At the global scale the urban system is closed as it comprises every city and all their links. The urban system within each country, however, has varying degrees of openness because inter-city links take place across national boundaries. In some countries the urban system is wide open because its cities have many and varied links with places elsewhere. In others, for reasons of geography, history and politics, the urban system is effectively closed to most external influences and operates in relative isolation.

The size distributional pattern is believed to reflect the number, direction and strength of the economic forces which act upon a country's cities. Primacy is held to be the simplest and most elementary arrangement in which a few simple forces act strongly upon a single centre. Such a situation is most common in developing economies with relatively closed or isolated urban systems where one place, commonly the national capital, is the focus of internal and international trade and so grows to dominance. At the other extreme, rank-size distributions are found where, because of the openness of the system, many forces affect the urban pattern in many ways. With greater economic complexity, there is a wider range of urban specialisms and more stimuli to urban development, all of which result in a rank-size pattern.

Smith (1985a, 1985b) suggests four reasons why primacy may exist in developing countries. The first is associated with colonialism and arises because empires tend to be controlled through key cities which, as foci of imperial interchange, operate at a level different from and higher than local or indigenous cities. Primacy is thus a function of colonial control, an explanation which appears adequately to explain the existence of dominant cities in Asia, but not in Latin America, where colonial rule ceased much earlier (Berry, 1971; Hay, 1977). Primate cities are seen in the second interpretation as the major outlets for the products generated in dependent export economies. They are the points of linkage between interior producing regions and external overseas markets. As such they benefit uniquely from processes of urban growth which are associated with trade. An important consequence of modern means of transportation, especially railways, is that they funnel export commodities from producing areas to the primate city, without contributing to the need for anything but very minor population centres in the hinterland, so creating a two-tier urban hierarchy and maintaining primacy. The importance of the railways in opening up interior areas of Latin America and in focusing development at the ports is well recognised, as Wilhelmy's work on Argentina shows (1986). The general applicability of export dependency explanations of urban primacy in the continent is confirmed by Smith (1985a), who found that in seven of the eight Latin American countries studied, the level of export production correlated with the degree of primacy.

Smith's third argument is that primacy may be created from within by the collapse or decline of the rural economy. Local industry and trade are often destroyed by export dependence, thus undermining the economic base of provincial centres. In this case, the largest city grows at the expense of the smallest. Primacy, finally, may be a social consequence of the transition of an economy from subsistence to capitalist production. Such a change typically transforms class and labour relationships and in particular leads to a reduction in the amount of labour which is required in agriculture. Those who are no longer needed in farming tend to concentrate in, and so inflate the size of, major cities where there are possibilities of jobs in service activities, or opportunities for income generation within the informal sector. Smith's research in Guatemala leads her to identify this process of labour release and urban concentration as the major reason for the continuation of primacy in parts of Latin America where colonialism, export dependency and rural collapse no longer apply (1985b). Such are the incidence and persistence of primacy in the developing world that they point to the operation of several processes working in combination. Primacy may not have a single simple explanation; rather its origins are more likely to be multi-causal.

A basic problem with the analysis and explanation of city-size distributions is that the size of many countries is poorly depicted by their population or geographical area. Indonesia, Japan and New Zealand consist of separate islands which must to some extent function in isolation, whereas the elongated shape of Chile, the Gambia and Laos inevitably affects the level of internal integration. Political boundaries tend to shift over time, so that it is difficult to relate urban size patterns to national historical and economic circumstances. The evidence suggests, however, that primacy exists in countries in which the principal city is more strongly linked to and integrated within the global urban system than it is to the domestic urban hierarchy. This situation is especially common in colonies, former colonies, and politically independent but economically dependent export economies.

Primacy points to the existence of a two-tier urban system in many of the countries of the world. People either live in one of the myriad of small villages or in the primate city: there are few settlements of intermediate size in between. Primacy can be regarded as an indicator of 'over-urbanisation', in the sense that the largest city has far outgrown all the others and has become an atypical and disadvantageous form of settlement in a particular country (Gugler, 1988). Other commentators focus upon the lack of middle-sized settlements and the implications for the way in which the urban hierarchy functions. A deficient middle tier is a well documented feature of the urban system in most African countries (Rondinelli, 1989). For example, in 1980, only Zambia, Zaire, Algeria, Egypt, Morocco and Nigeria had more than five cities, other than the largest urban centre, with a population greater than 100,000. Four of these countries accounted for nearly 70 per cent of Africa's secondary cities. Eleven other countries had only one or two secondary cities and only in Zambia, Nigeria and Morocco did all of the secondary cities combined have a larger percentage of the urban population than the biggest city. Close parallels exist in India, although here the major shortfall is in the number of market towns which provide points of articulation between village and the urban systems (Ramachandaran, 1993). Differences of size are inevitably reflected in an imbalance of importance and role. Primate cities typically dominate their countries in economic and political terms. Invariably they are the centres in which national elites and other major decision-makers and opinion leaders are concentrated. They are the headquarters for national television and telephone services and have the principal, indeed probably the only, international airport. In turn, they are the link points through which the dependent village population is connected to the global urban system.

Studies of city-size distributions provide an indirect insight into the characteristics of urban linkages and connectivity. Together with work on the distribution of the urban population they paint a picture of gross unevenness and variable integration. Not only are the spread of urban population and the level of urban development far from uniform, but cities differ significantly in the extent and ways in which they interconnect and interdepend. Cities in countries with rank-size patterns tend to be well integrated; those in countries with primate distributions, with the exception of the primate city itself, are predominantly inward-looking and have their strongest connections with the indigenous economy. These differences suggest that the global urban system is presently fragmented and incomplete. Rather than a coherent whole, the contemporary urban world consists of a set of loosely-knit sub-systems. The largest is global in extent and is based upon movements and exchanges of people, goods, images, information and ideas among rank-sized urban centres and primate cities. It is dominated by the cities in the core economies but also includes the primate and principal centres in the periphery. Other sub-systems, in the periphery, have primate cities as their apexes but have few external links except through the primate city and so function primarily at a local scale. Still others, in the most remote and backward parts of the periphery, are divorced from national urban systems and so operate in comparative independence and isolation.

THEORIES OF URBAN FORMATION

The world is an urban place because towns and cities offer substantial benefits over other forms of settlement. The advantages which people derive from clustering together are greater than when they scatter and disperse. Some of the most fundamental questions in urban geographical study are concerned with the precise nature and power of these agglomerative tendencies, although Wheatley (1971: p. 318), among others, suggests that 'it is doubtful if a single autonomous causal factor can be identified in the nexus of social, economic and political transformations which result in urban forms'. Theories of urban formation seek to identify the forces which permit and encourage large numbers of people to concentrate in comparatively small areas in space. As such they form the bases of the theories of urban growth and urbanisation which are discussed in the next chapter. Two broadly contrasting viewpoints are prevalent in the literature, one underlining the primacy of economic benefits, the other emphasising the roles of the social bond.

Economic interpretations of urban formation lay stress on the savings of assembly, production and distribution costs which may be achieved through concentration. They argue that the existence of towns and cities is a consequence of the search for the most economical form of settlement. In primitive economies based upon labour intensive agriculture, the population is typically arranged in small village communities which are scattered across the countryside, a pattern which gives maximum access to the land. Towns and cities, according to this interpretation, come into existence when the level of agricultural production generates an annual food surplus (Childe, 1950). This critical development frees part of the workforce from agricultural employment and makes possible a range of craft and trade activities which cluster together in space so as to gain the maximum benefits of economies of scale and agglomeration. Cities are thus formed through the geographical concentration of social surplus product (Harvey, 1973: p. 216). Such activities are the core elements in the urban economy.

Underlying this economic interpretation of urban formation is a set of relationships which are best explained by economic base theory. At its most simple level, the urban economy may be viewed as two interdependent sectors, the basic and the non-basic (Figure 7). Cities can only exist at the expense of rural areas by buying in food and most raw material and product requirements, and so any importing activity is 'city-forming' in that it makes possible and sustains the urban population. The basic sector consists of all those activities and employment which produce goods and services which are sold outside the city and provide the finance to enable basic requirements to be imported into the city. Corn and seed merchants, agricultural advisory services and farm machinery manufacturers who are urban-based and who serve a non-urban market are obvious examples of basic activities, but the classification also includes a wide range of manufacturers and service providers who 'export' their products across the city boundary. Although these city-forming functions are concerned exclusively with external markets, they themselves generate demands for goods and services for their own support within the city. The non-basic sector consists of all those activities which provide goods and services for the city itself. Examples of 'city-serving' activities include municipal government; street cleaning services; police, fire and ambulance services; corner stores; video hire shops and take-away food outlets. Together the basic and non-basic sectors account for all the activities and employment in the city.

The two sectors are functionally interdependent. Any change in the size of one sector will be associated with a change in the size of the other. If the basic sector expands, workers in that sector will spend more on city services, so the non-basic sector will grow as well. Differences in the size of the two sectors, however, mean that changes in one will have a differential effect upon the other. For example, in a city with a basic:non-basic ratio of 1:3, an increase of 10 in basic employment will yield an increase of 40 (10 + 30) in total employment. The urban economic multiplier is a central

32

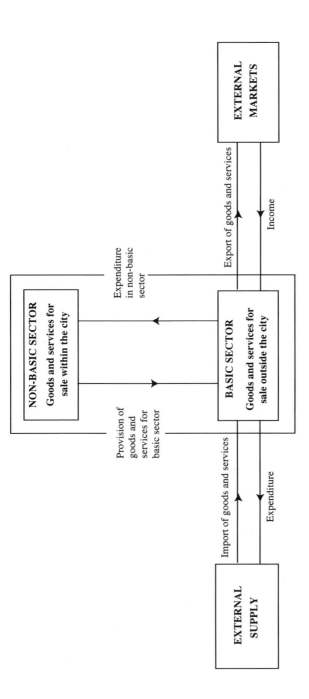

Figure 7 Elementary structure of the urban economy

➤ economic flows

EXTERNAL MARKETS

Export of goods and services

Income

NON-BASIC SECTOR
Goods and services for
sale within the city

Expenditure
in non-basic
sector

BASIC SECTOR
Goods and services for
sale outside the city

Provision of
goods and
services for
basic sector

Import of goods and services

Expenditure

EXTERNAL SUPPLY

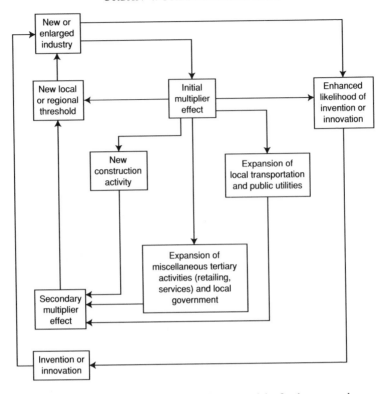

Figure 8 The circular and cumulative model of urban growth
Source: adapted from Pred (1977)

concept in the explanation of urban formation and growth. It represents a mechanism whereby increases in the volume of external trade,
and hence the size of the basic sector, result in a corresponding
growth in employment in the non-basic sector, and thus an overall
increase in employment and population in the city as a whole. In
practice, the mechanisms of growth are of course highly complex.
For example, if business output expands in response to increased
sales, so profits and wages will increase, expenditure by workers and
shareholders will rise, the demand for labour will grow, and the
population will reach new levels or thresholds, resulting in the entry
of new firms into the market (Figure 8). Urban growth is best seen
as a circular and cumulative process, as growth in one sector trig-
gers expansion through secondary multiplier effects elsewhere in the
economy (Pred, 1977).

34

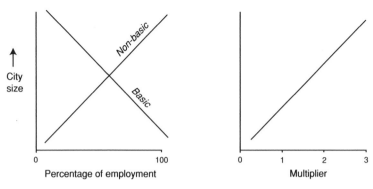

Figure 9 General relationships between basic and non-basic components, and the urban economic multiplier, with city size

Studies of the economic structure of cities are difficult because many people's jobs involve both 'exporting' and 'city-serving' activities, so the analytical procedures which have been developed produce only crude breakdowns. The most widely used is the 'minimum requirements approach', as pioneered by Ullman and Dacey (1962). This involves classifying cities into size groups and then examining the percentage of the total labour force which is employed in each major category of occupation. The lowest percentage recorded for any city in each group is assumed to be the minimum necessary to enable cities of that size to function. These minimum requirements are equated with the non-basic workers, and the number of workers over this minimum figure is taken to represent the basic labour force. Empirical studies using this and related methods have identified the relative size of the two sectors. They show that the basic sector decreases in size as the urban population increases (Figure 9). For example, in a city of 10,000 people, approximately two-thirds of all employment is in basic activities, whereas in a city of 15 million, the figure is nearer one-quarter. A related finding is that the size of the multiplier similarly increases with urban size. It has a value of around 0.75 for a city of 200,000, but is nearer 2.00 for a city of six million.

Several important theoretical and practical implications follow from these findings. The first is that the larger the city, the less it is dependent upon basic activities, and hence its links with surrounding suppliers and markets, for its viability. Whereas the volume of external trade is critical for small towns and determines whether they expand or decline, it is of secondary importance for the metropolis. Beyond a critical size, thought to be in the region of 250,000, growth is self-generated and is a product of the non-basic sector. The second consideration, which relates to the size of the urban multiplier, is that larger urban centres have the capacity to grow more rapidly: a small increase in the basic sector leads to proportionately large increases in the non-basic sector. Conversely, this means that large cities are somewhat more vulnerable to collapse if the basic sector suddenly contracts. These mechanisms and relationships emphasise the close dependence of small cities on the volume of surplus product that exists locally. They provide theoretical support for the views of Childe (1950) that the generation of surpluses was a key factor in urban genesis.

It is important to emphasise that cities exist because of and at the expense of their surrounding environments. This applies to imports of food surpluses as summarised by elementary economic base theory, and, when their total environmental needs are considered, to their dependence on external sources of air, water and energy as well. For Odum (1989), the underlying relationship is fundamentally parasitic, since cities make no food, clean no air and clean very little water to a point where it can be reused. An important practical consideration is the long-term viability of these relationships and its consequence for a sustainable urban future, issues which are considered in more detail in Chapter 8. Many authors are deeply pessimistic. One view of cities is that they are overgrown and have fast-declining carrying capacities, so that only catastrophe awaits. Similar opinions are voiced by Friedman (1984: p. 48) who sees the city as a cancer, 'an overgrown organ which takes all the food, so much food that it can no longer perform its proper function; and cancer is a lethal illness'.

Although urban places need to import agricultural surpluses, it is by no means certain that this was the formative event in the first emergence of cities: 'It has not yet been demonstrated clearly and unequivocally first, that a generalised desire for exchange is capable of concentrating political and social power to the extent attested by the archaeological record, or second, that it can bring about the institutionalisation of such power' (Wheatley, 1971: p. 282). The fact that urban centres are present in a wide range of economies and cultures throughout the world suggests that origins of urban living are a product of human relationships and lie in the interpersonal ties which encourage people to congregate in space. Social explanations of urban formation stress the gregarious nature of human behaviour. They point to the complementary properties of links such as male and female, mother and child, sender and receiver, speaker and listener, giver and taker, and argue that such bonds introduce strong centripetal tendencies among human populations. Even small groupings offer security, defence, self-help and the prospects of finding a mate, so they are attractive to non-group members. With increased membership, the value of community benefits grows. Cities emerge when social institutions and mechanisms, in the form of defence, administration, government and religion, are developed which enable the population to live together in sizeable concentrations in space. Social organisation is, therefore, accorded priority over economic developments as the independent variable in urban formation.

These arguments are most closely associated with the work of Adams (1966) on the emergence of cities in early Mesopotamia. He maintained that the rise of cities was pre-eminently a social process, an expression more of changes in people's interactions with each other than with their environment. The novelty of the city consisted of a whole series of new institutions and a vastly greater size and complexity of social unit, rather than basic innovations in subsistence. For Lampard (1965), society has evolved through a number of organisational stages, each of which is associated with different settlement forms. Particular emphasis is placed upon the 'primordial', since this represents the achievement of a level of organisation which is necessary to support and sustain village life. Improvements in agricultural productivity are an essential prerequisite, but the development of stable communities, capable of elementary agricultural management and a degree of social control, is crucial to the viability of the settlement. The existence of cities, the next stage, represents the achievement of a higher level of social sophistication and consensus. This is reflected in formal bureaucratic, religious, military and political systems, which together enable large concentrations of people to coexist peacefully, harmoniously and prosperously. Underlying the argument is a strong emphasis on the urban consequences of social development. While the role of economic productivity in the rise of cities is not denied, its importance is held to be contingent upon the achievement of a given stage in social development. In this way, the city is 'a mode of social organisation which furthers efficiency in economic activity' (Lampard, 1955: p. 92).

Although these ideas on urban formation were developed to account for the rise of the first cities, their theoretical underpinnings and their implications are of substantial relevance today. The development of appropriate social structures is as necessary for ensuring the stability and prosperity of contemporary mega-cities as it was to the viability of early agricultural settlements. The nature of urban economic bases and multiplier effects are issues of central importance in planning and urban development. Cities exist on the basis of surplus product, whether food supplies, which were of critical significance in antiquity, or manufactured goods and services, which are the principal items of modern economic exchange. The ways in which such flows focus upon individual centres are primary determinants of the structure and organisation of national urban hierarchies and the arrangement of centres into a world system of cities.

CONCLUSION

This chapter has been concerned with basic distributions and elementary explanation: how many urban people there are, where they live, why they live there, the ways in which their cities interact, and the implications for economic and social development. Attempts to overview patterns at the world scale, using nation states as the basic units of measurement, are difficult because of the underlying complexity so there is inevitably a very fine line between useful generalisation and naive over-simplification. They are not helped by the wide variations which exist in the size of countries and in the availability and quality of their census data. What the statistics show, however, is that the world is now more urban than rural and that urban development is well established in most countries and regions. Many people live in the towns and cities of China and India, but urban populations are low throughout most of Africa and much of the rest of Asia. Levels of urban development are highest in South America.

Despite the importance of million and mega-cities, especially in countries with primate urban patterns, most people live in small or intermediate centres whose size relates closely to the productivity and sophistication of their local economies. Such evidence as can be drawn from studies which seek to explain city-size distributions suggests that although the present level of urbanisation is in excess of 50 per cent, the degree of global urban integration remains low. The concept of the global urban system has considerable relevance to settlements in core areas, primate cities and their closely dependent intermediate centres, but many of the smaller places in the periphery function in comparative independence and isolation.

The urban world is the product of processes of population concentration that began many millennia ago when the generation of annual food surpluses and the creation of viable social structures first led to and enabled the establishment of towns and cities. Although the underlying social and economic prerequisites are the same, the patterns outlined and explained in this chapter owe most to recent and continuing trends which operate at a global scale. It is only in the last half-century that urban development has become sufficiently widespread across the world as to suggest causes which relate to the emergence of the world-economy. Having taken a snapshot view of the contemporary urban system, it is now appropriate to explore the dynamics of global urban change by analysing processes and patterns of urban growth and urbanisation in detail.

RECOMMENDED READING

Brunn, S. D. and Wheeler, J. F. (1993) *Cities of the World*, London: Harper Collins.

A set of specially commissioned chapters which examine levels of contemporary urban development on a region-by-region basis. Particular attention is paid to the principal cities in each area.

Carter, H. (1995) *The Study of Urban Geography*, (4th edn) London: Arnold.

A standard text in urban geography which includes a useful discussion of the economic and social interpretations and explanations of the origins of towns and cities.

Dogan, M. and Kasarda, J. (1989) *The Metropolis Era: Vol. 1 A World of Giant Cities; Vol. 2 Mega-Cities*, London: Sage.

Volume 1 examines the characteristics and processes of urban growth in both the developed and the developing worlds. Volume 2 presents an in-depth analysis of ten giant cities around the world.

Sit, V. F. S. (1985) *Chinese Cities: The Growth of the Metropolis Since 1949*, Oxford: Oxford University Press.

A collection of essays by British and Chinese geographers about data sources and the major urban centres.

Timberlake, M. (1985) *Urbanisation in the World-Economy*, London: Academic Press.

A set of edited papers on different aspects of urban development at the global scale, including especially useful contributions by C. A. Smith on city-size distributions and their explanation.

3

URBAN GROWTH
AND URBANISATION:
HISTORICAL PATTERNS

The present pattern of global urban development is merely the most recent product of processes of urban change that began over eight thousand years ago. It represents an intermediate stage in the progression from a wholly rural to what will possibly be a completely urban world. The global urban pattern is changing in three different and unconnected ways: through urban growth, urbanisation, and the spread of urbanism. Urban growth occurs when the population of towns and cities rises. Urbanisation refers to the increase in the proportion of the population that lives in towns and cities. Urbanism is the term which is most commonly used to describe the social and behavioural characteristics of urban living which are being extended across society as a whole as people adopt urban values, expectations and lifestyles. This chapter and Chapter 4 identify and attempt to account for recent patterns of urban growth and urbanisation at the global scale. The origins and spread of urbanism are the focus of Chapters 5 and 6.

Urban growth and urbanisation are separate and independent trends. Urban growth refers to the absolute increase in the size of the urban population. It occurs both through natural increase, that is an excess of births over deaths, and through net in-migration. In most cities both factors operate together and reinforce each other, although the relative balance may vary from place to place and at different times. Growth rates are compounded when the in-migrants are young adults. These are the most fertile age group and their influx is likely to raise the rate of natural increase. However if the in-migrants are predominantly of one sex, the accompanying rate of natural increase is likely to be lower. As an actual rather than a percentage figure, urban growth is not subject to any ceiling. It can take place without urbanisation occurring so long as rural growth occurs at the same rate. It is likely to continue after urbanisation has ceased as the population, which will all be living in urban places, goes on growing through the excess of births over deaths.

Urbanisation measures the switch from a spread-out pattern of human settlement to one in which the population is concentrated in urban centres. It is concerned with the relative shift in the distribution of population from the countryside into the towns and cities. Urbanisation is a change that has a beginning and an end, the former being when the population is wholly rural, the latter occurring when everyone is recorded as living in an urban place. In a world in which half the population presently resides in urban places, this latter situation is clearly some way off and is unlikely ever to be achieved. A sizeable rural population engaged in agriculture, fishing and forestry will always be required. There is growing evidence from contemporary experiences in developed countries that urbanisation tends to rise to a peak of around 85 per cent urban, and then falls slightly. Models of urban spatial evolution suggest that levels of urban development of around 80 per cent are probably the optimum for sustainability.

It is important to emphasise that as the total population of a country consists of both urban and rural dwellers, an increase in the 'proportion urban' is a function of both. It occurs when the urban component increases in relative size, either through faster urban growth or more rapid rural population decline. The measurement of urbanisation is not without its difficulties as it depends upon the division of a country into urban and rural areas. It is affected by changes in definition and the classification of centres which are made from time to time by national census authorities. Such technical issues are addressed in the Appendix.

URBAN GROWTH

By far the most important characteristic of contemporary urban change is the sheer scale of urban population growth. Each year between 1985 and 1995 the world's urban population rose by a staggering 73 million. Urban growth correlates strongly with overall population growth and so it is not surprising to find that the greatest gains occurred in highly populated countries where large numbers are added to the national population each year (Figure 10). The urban population of China alone rose by 226 million over the decade. This vast increase in only ten years is almost as large as the present

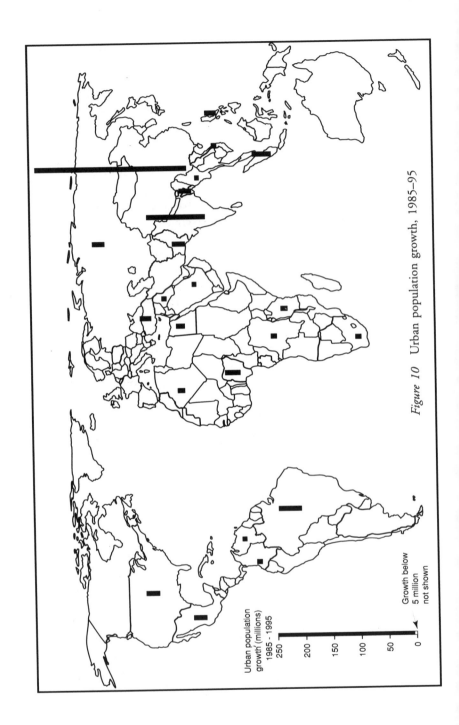

Urban population growth
1985 - 1995

250

200

150

100

50

0

Growth below
5 million
not shown

Figure 10 Urban population growth, 1985–95

population of all the towns and cities of South America. Major increases also occurred in India (87 million), Brazil (32 million), Indonesia (26 million) and Nigeria (22 million). Little or no urban growth took place in Europe, where national population levels are virtually static. For example, the urban population of Germany rose by little more than one million between 1985 and 1995.

Although the scale of increase is greatest in China, urban populations are growing rapidly throughout Africa and southern Asia (Figure 11). The rates are highest in parts of Sub-Saharan Africa and the Middle East. Afghanistan leads the way with an annual average rate of growth of urban population, between 1990 and 1995, of 8.5 per cent, and high rates of growth in the same region also occurred in Oman and Yemen. The countries of Sub-Saharan Africa are predominantly rural and have few sizeable cities, but they are also gaining urban populations rapidly. Botswana, Swaziland, Mozambique, Rwanda and Tanzania all had rates of urban growth in excess of 7 per cent over the period.

Urban growth in most countries is a product of both high rates of natural increase of the urban population and net in-migration. Despite the attention which is commonly paid to migration, there is little difference in their relative importance and the two processes compound and reinforce each other. The relative contribution of the components of urban growth is shown in an analysis of urban demographic change in twenty-nine countries reported by Preston (1988). He found that in twenty-four cases, the rate of natural increase of the population in cities exceeded that of net in-migration. The mean percentage of urban growth which was attributable to natural increase was 60.7 per cent. In India it was 61 per cent during the period 1951–61, 65 per cent between 1961 and 1971, and 53 per cent during 1971–81 (Bradnock, 1984).

More recent data, for twenty-four developing countries, reported by Findley (1993; p. 15) show that the average migrant share of urban growth over the period 1975 to 1990 was 54 per cent. Findley points out, however, that this figure underestimates the true picture because many migrants who live and work in the city do not move there permanently. A large number of migrants in the cities of the developing world circulate between urban and rural areas in repeated movements, some staying for a few months while others remain for several years. Such research as is published on the topic indicates that temporary or circular migrants are most common in South East Asia and West Africa, where they make up between 33 and 70 per

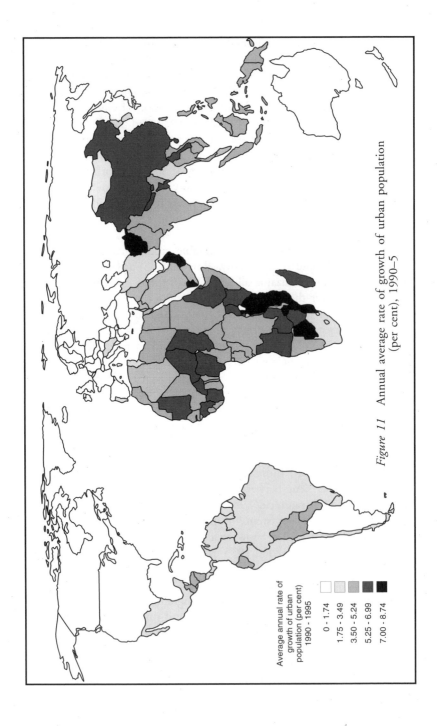

Figure 11 Annual average rate of growth of urban population (per cent), 1990–5

Average annual rate of
growth of urban
population (per cent)
1990 - 1995

0 - 1.74
1.75 - 3.49
3.50 - 5.24
5.25 - 6.99
7.00 - 8.74

cent of the total migrant population. On this basis, official statistics may miss one to two out of every three migrants.

Natural increase is an important component in urban growth, despite the fact that both birth rates and death rates in cities are normally lower than in rural areas. One reason for the lower birth rate is the substantial number of young male migrants in the city who are unmarried or who have left their wives at home in their village. Another is that artificial birth control methods are more widely available in urban areas. At first glance the lower death rate seems surprising, given the poverty and squalor of many cities, but it can partly be explained by age structure. Young adults in every society are the fittest group and their presence in such large numbers in most cities reduces the overall death rate. Urban places in the developing world, however, are generally more healthy than their rural hinterlands. They benefit from investment in drinking water and sanitation and are places where medical and scientific techniques, expert personnel, and funds from the developed nations are first imported and where most people can be reached at least cost. A specific factor is that more births tend to take place in hospital and so rates of perinatal and infant mortality are lower. Diseases such as malaria and cholera can more easily be controlled or treated in cities than in the countryside.

Little or no growth is occurring in urban populations throughout most of the developed world. Growth rates of less than 0.5 per cent per annum were recorded over the 1990–95 period in the United Kingdom, Belgium, Germany, Italy, Denmark, Sweden and Finland. Such low overall rates of growth are a product of trends which are affecting cities of different size and do not mean that the urban hierarchy is unchanging. Generally, metropolitan centres are losing population while towns and small cities are gaining. The most significant of these trends is the fall in the population of major cities. Urban population decline in developed countries is very much a feature of the past twenty years. It was first identified by Berry in 1976 and was confirmed by the most recent censuses for the USA and the countries of north western Europe (Clark, 1989). Nineteen major cities in the US manufacturing belt lost population during the 1980s, as did twenty-two of the largest two dozen cities in the United Kingdom. Sizeable losses were similarly incurred by the metropolitan centres of north western Europe. A slight overall increase in the urban population of these countries, however, took place because of a compensatory rise in the numbers of people living in small and medium-size towns and cities.

Table 2 Urban agglomerations with eight million or more persons, 1950–2000

1950	1970	1990	2000
More developed regions			
New York	New York	Tokyo	Tokyo
London	London	New York	New York
	Tokyo	Los Angeles	Los Angeles
	Los Angeles	Moscow	Moscow
	Paris	Osaka	Osaka
		Paris	Paris
Less developed regions			
None	Shanghai	Mexico City	Mexico City
	Mexico City	São Paulo	São Paulo
	Buenos Aires	Shanghai	Shanghai
	Beijing	Calcutta	Calcutta
	São Paulo	Buenos Aires	Bombay
		Bombay	Beijing
		Seoul	Jakarta
		Beijing	Delhi
		Rio de Janerio	Buenos Aires
		Tianjin	Lagos
		Jakarta	Tianjin
		Cairo	Seoul
		Delhi	Rio de Janerio
		Manila	Dhaka
			Cairo
			Manila
			Karachi
			Bangkok
			Istanbul
			Teheran
			Bangalore
			Lima

Source: United Nations (1991) *World Urbanization Prospects*, Table 11.

An important corollary of contemporary urban growth at the global scale is the rapid increase in the number and size of the largest cities. Against the background of a general rise in the number of people who live in urban places it is the metropolitan centres that are proliferating and growing the fastest. United Nations estimates indicate that the number of cities with over 8 million people increased from 10 in 1970 to 20 in 1990 (Table 2). Those in the size range

2 to 4 million went up from 39 to 72 (Dogan and Kasarda, 1989: p. 13). This growth is indicative of an emerging pattern of exceptional population concentration. The urban world is increasingly and quickly becoming a world of mega-cities.

A remarkable feature of contemporary urban growth is that the number and size of mega-cities are increasing most rapidly in developing countries. In 1950, the only mega-cities, London and New York, were both in the developed world (Table 2). By the year 2000 there are expected to be twenty-eight, of which twenty-two will be in the developing world. The annual average growth rates of all the emerging mega-cities in the developing world is expected to be some 3.5 per cent. It is likely to be especially high in Dhaka (7.0 per cent), Lagos (5.6 per cent), Bangalore (5.7 per cent), Delhi (4.6 per cent), Jakarta (4.4 per cent), Karachi (4.4 per cent), Istanbul (4.1 per cent) and Bangkok (4.1 per cent). Mega-cities will be found in all regions by 2000 although there will only be two, Cairo and Lagos, in Africa. Many of these mega-cities will be extremely large, with Bombay, Calcutta, Mexico City, São Paulo, Shanghai and Tokyo, as well as New York, having populations in excess of 15 million people. Others will be large in relative terms as they will house a major share of their country's total urban population. This is expected to be in excess of 30 per cent in the case of Cairo, Buenos Aires, Lima, Bangkok, Dhaka and Seoul. Such is the rate of proliferation and growth of mega-cities that, by the end of the century, they are expected to house some 12 per cent of the world's urban population.

URBANISATION

The proportion as well as the total number of people who live in towns and cities is also increasing at the global scale. Urbanisation involves a significant shift in the distribution of population from rural to urban locations. Each year some 312 million more people are added to the world's towns and cities than to its rural areas. Although at 0.54 per cent per annum the global rate of urbanisation seems comparatively modest, it has profound implications for the long-term distribution of population. Around 25 per cent of the world's population lived in towns and cities in 1950. It is likely to be 75 per cent by 2010 (Figure 12, insert).

Urbanisation is a cyclical process through which nations pass as they evolve from agrarian to industrial societies. For Davis (1969) the

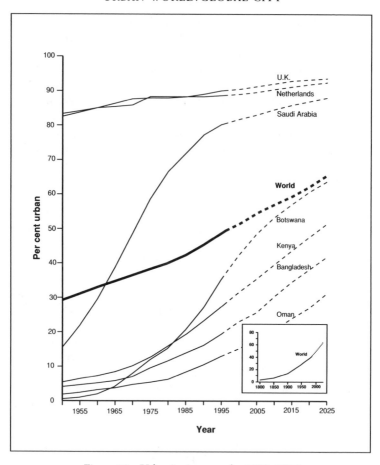

Figure 12 Urbanisation trends, 1950–2025

typical course of urbanisation in a given territory can be represented by a curve in the form of an attenuated 'S' (Figure 12). Such a curve, which is characteristic of many growth processes over time, is known as a logistic curve. The first bend in the curve is associated with very high rates of urbanisation as a large shift takes place from the country to towns and cities in response to the creation of an urban economy. It is followed by a long period of consistent moderate urbanisation. As the proportion climbs above about 60 per cent the curve begins to flatten out, reaching a ceiling of around 75 per cent. This is the level at which rural and urban populations appear to

achieve a functional balance. Historical data suggest that the highest
rate of global urbanisation occurred at the end of the last century,
when the distribution of the population in what were then the leading industrial economies was switching rapidly from rural to urban.
The rate at present is moderate and is a consequence of the progressive urbanisation of the population in developing countries.

At any one time, individual countries are at different stages in
the cycle so it is necessary, in making sense of present rates of urbanisation (Figure 12), to take account also of the overall level of
urbanisation (Figure 4). Some countries, such as Saudi Arabia and
Botswana, are presently experiencing high rates of urbanisation as
their populations switch rapidly from rural to urban locations. In
others the rate is negligible, either because the cycle of urbanisation
is complete, as in the United Kingdom and the Netherlands, or
because, as in Oman and Bangladesh, it is only just beginning.
Further substantial shifts in the distribution of population can be
expected in many countries with presently low levels of urbanisation
as the proportion of the population that lives in towns and cities
rises to ceiling levels.

The switch in the location of population from rural to urban is
presently quickest in many of the countries of Africa and Asia. It is
most marked in East Africa, where the annual average rate over the
period 1990 to 1995 in Botswana, Mozambique, Tanzania and
Rwanda was in excess of 4.0 per cent. It is also high in Nepal (4.1
per cent) and China (3.9 per cent). These are the countries which
are going through the high rate phase of the logistic cycle and where
the social and economic tensions associated with urbanisation are,
in consequence, most severe. Elsewhere in Africa and Asia the rate
is lower because the cycle of urbanisation has only recently started.
Countries with annual average rates of urbanisation below 2.0 per
cent and levels of urbanisation below 20 per cent include Cambodia,
Afghanistan and Sri Lanka.

Urbanisation is presently a developing world phenomenon. The
highest rates are to be found in the less developed countries, while
little urbanisation is occurring in the already highly urbanised developed world. An important exception to this gross generalisation is
the countries of South America where urbanisation has effectively
ceased but where levels of economic development are low. In examining the overall association between urbanisation and development,
the World Bank estimates that urbanisation is increasing three times
faster in low and middle income countries than it is in high income

countries. Contemporary urbanisation involves the large-scale redistribution of people in many of the world's poorest nations which are least able to cope with its consequences. Because the rate is greater, levels of urban development in the developing world are catching up with and will soon approximate those in the developed world.

The geography of contemporary urbanisation is similar to but is not the same as the geography of urban growth. Countries which are presently urbanising most rapidly are mostly those in which there are the highest rates of urban growth (Figures 11 and 13). Rates of both urbanisation and urban growth are presently highest in East Africa. The settlement patterns in this region are being transformed most radically through the combined effects of urban population increase and redistribution.

Little change is taking place in the urban and rural balance in the developed world because, in most countries, the cycle of urbanisation has run its course. More detailed analysis in fact shows that in many developed countries, the processes responsible for urbanisation have turned around. After many decades of relative decline, the rural component of the population is now increasing relative to the urban component. The net effect in geographical terms is that there is a shift of population at the national scale from a state of more concentration to one of less concentration. For Berry (1976) this amounts to counterurbanisation, insofar as the traditional processes which favour towns and cities at the expense of rural areas are now working in reverse.

Counterurbanisation replaced urbanisation as the dominant process of locational change in the United States more than two decades ago. Between 1960 and 1970 the metropolitan areas of the United States grew five times as fast as the rural areas. But during the 1970s the pattern was inverted, with the rural areas gaining population at one-and-a-half times the rate of that in the cities. A breakdown of the data reveals the extent to which this change is attributable to counterurbanisation. The work of Berry (1976) showed that approximately half of the new non-metropolitan growth was adjacent to cities and amounted to no more than suburban sprawl across excessively tightly drawn boundaries. A roughly equal portion of non-metropolitan growth was, however, non-adjacent to and remote from existing cities and so represented true non-metropolitan revival. More detail is provided by an analysis of the changing metropolitan structure of northern Ohio between 1960 and 1970 (Berry and

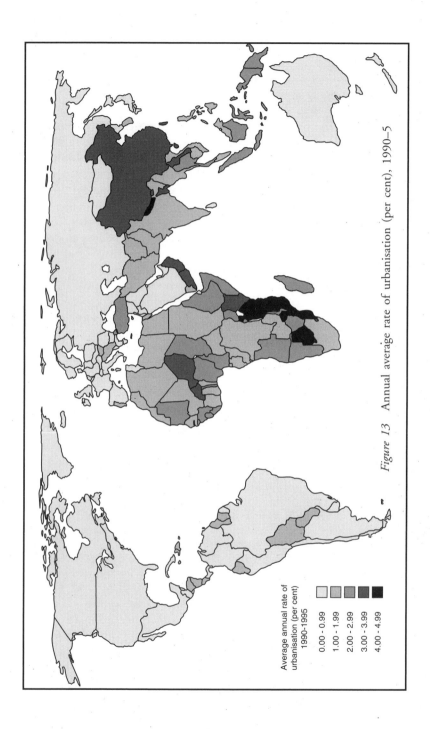

Figure 13 Annual average rate of urbanisation (per cent), 1990–5

Average annual rate of
urbanisation (per cent)
1990-1995

0.00 - 0.99
1.00 - 1.99
2.00 - 2.99
3.00 - 3.99
4.00 - 4.99

Gillard, 1976). Maps showing commuting to central Cleveland and central Akron in the two years identified significant population growth in inter-metropolitan areas. The lifestyles of people living in these expanding areas owed little to the daily influence of the city. Such growth was the product of counterurbanisation insofar as it had arisen in areas remote from, and with no effective regular ties with, the traditional urban core.

Research workers have since confirmed that counterurbanisation is widespread throughout the developed world and as such represents a distinctive and advanced stage of urban development. In their massive survey of growth centres in the European urban system, Hall and Hay (1978: p. 225) reached a general conclusion that 'at least down to 1970, Europe did not demonstrate any tendency to deconcentration'. By 1982, however, clear evidence existed that urban decline was well established in the Atlantic region (Great Britain and Ireland), and in parts of northern, western and central Europe. However, Spain, Portugal and Italy were characterised by continuing urban growth. A similar conclusion was reached by Fielding (1982). In nine out of the fourteen western European countries studied, he found that urban population growth ceased between 1950 and 1980. In seven of these countries the principal cities were, by the end of the period, in decline while rural and small towns were gaining in population. The prevalence of counterurbanisation within contemporary processes of urban change in the United Kingdom is further emphasised in the work of Champion (1989) and Cross (1990). Both show that the population is progressively deconcentrating as the number of people who live in rural areas rises in relation to those in the cities.

STAGES OF URBAN DEVELOPMENT

If the Western metropolis is the most advanced form of settlement, then counterurbanisation may represent the achievement of a distinctive phase in the urban life-cycle which will be followed by cities in what is presently the developing world. For Berg *et al.* (1982), expanding on the ideas of Hall *et al.* (1973), cities evolve in a clearly defined sequence of stages which can be conceptualised in a model of urban development based upon population changes in urban regions as a whole and upon shifts of population within urban regions. A modified version of Berg's model is shown in Table 3. A fall in urban populations is most characteristic of the counterurbanisation stage

Table 3 Stages of development of a daily urban system (DUS)

Stage of development	Classification type	Population change characteristics			
		Core	Ring	DUS	
I Urbanisation	1 Absolute centralisation	+ +	–	+	Total growth (Concentration)
	2 Relative centralisation	+ +	+	+ + +	
II Exurbanisation	3 Relative decentralisation	+	+ +	+ + +	
	4 Absolute decentralisation	–	+ +	+	
III Counterurbanisation	5 Absolute decentralisation	– –	+	–	Total decline (Deconcentration)
	6 Relative decentralisation	– –	–	– – –	
IV Reurbanisation	7 Relative centralisation	–	– –	– – –	
	8 Absolute centralisation	+	– –	–	

Source: modified from Berg *et al.* (1982: 36)

Note: The terms urbanisation, exurbanisation, counterurbanisation, and reurbanisation are defined as follows: Urbanisation occurs when the growth of the core dominates that of the ring, while DUS as whole is growing. Exurbanisation occurs when the growth of ring dominates that of the core, while the DUS is still growing. Counterurbanisation occurs when the growth of the ring dominates that of the core, while the DUS declines. Reurbanisation occurs when the growth of the core dominates that of the ring, while the DUS declines.

which follows periods of urbanisation and exurbanisation. It may in turn be succeeded by an era of reurbanisation, when cities return to a state of growth and expansion.

In the 'stages of development' model, changes of urban form are related to shifts in the distribution of population within and around the city. For this purpose it is useful to divide the urban landscape into a number of areas according to their population, employment and commuting characteristics. At the centre of the urban region is a core area consisting of a major concentration of population, jobs and economic activity. Within the core may be distinguished a central business district of shops and offices, surrounded by an inner area of mixed industrial, wholesaling, warehousing and residential uses which together constitute the central city. This area in turn is typically surrounded by a set of residential suburbs consisting, in outward progression, of estates and developments of successively newer and less densely packed houses. Together these morphological zones define the physical city.

Beyond the physically built-up area, which is commonly constrained by planning policies, is an extensive commuting area from which the city draws many of its daily workers. It encompasses an area of towns and villages in a predominantly rural setting in which the population focus their activities upon the core. The strength of the diurnal ties between the core and the ring means that the two areas together function as a daily urban system. They are tightly bound together in a relationship of interdependency by morning and evening commuting flows and by movements between the two areas for shopping and recreation. Beyond the daily urban system is an extensive but sparsely populated rural area. As a relatively self-contained labour market this region has no major urban centre of its own and so looks to the core on an infrequent basis and for only the highest order services.

Berg's model of urban development is based upon variations in the direction and rate of population change between the core and the ring. Two types of change are recognised. Shifts are absolute when the directions of population change in the two areas are different as, for example, when the core is growing while the ring is declining. Alternatively, the shift is relative when each area has the same direction of change but the rate of change is different. Thus a relative shift to the core would occur when both the core and the ring are growing but the population of the core is increasing at a faster rate. It is important to stress that the model is purely descriptive and makes no inferences as to how population shifts occur. Shifts of population arise because of differences between areas in the numbers of births, deaths and movements that take place. The relationships among these variables are typically so complex that simple associations, as for example between counterurbanisation and net out-migration, are likely to be misleading. Detailed analysis of the individual components of demographic change is necessary before reasons for observed population shifts can be advanced.

The urbanisation phase of urban development is the first of two stages which are characterised by the overall growth of the daily urban system (Table 3). It is associated with the expansion of employment opportunities in the city and with increases in the efficiency of agriculture which release workers from the land. The growth in the population of the daily urban system arises primarily because the core expands at the expense of the ring. Initially the process of centralisation of population within the daily urban system is absolute as the ring, which is still overwhelmingly rural, loses population to the core. It gives rise to a compact and densely populated physical city, as existed in Britain in the early nineteenth century and as is common in many parts of the developing world today. Later, as transport improvements allow some separation of place of residence from place of work, and as population spills over a tight city boundary, growth occurs in the ring. However, as this is less than that which is taking place in the core, continued urbanisation is the product of relative centralisation.

The most important feature of the 'exurbanisation' phase in the spatial evolution of the city is that the population decentralises. Although the daily urban system as a whole continues to increase in population at the expense of the surrounding commuter belt, differences in the rate, and subsequently in the direction, of change between the core and the ring mean that the daily urban system decentralises as it grows. The expansion of the ring reflects the environmental attractions of exurban locations and an increase in the number of people who can afford to move out of the core. It is facilitated by major transport improvements, including the development of suburban rail networks and the introduction of tram and bus services. Continued, though much slower, growth of the core means that a relative decentralisation characterises the initial phase of exurbanisation. Subsequently, as the core begins to lose population, the process of change becomes absolute. Core area decline and ring growth in the latter period are closely linked to rising levels of car ownership and use which enable large numbers of people to populate the commuter belt, especially those areas which are well away from the major radial routeways.

A decline in the population of the daily urban system, both core and ring together, distinguishes the third and fourth stages of urban development. In place of urban expansion, it is in the rural areas beyond the daily commuting range of the core that growth takes place. The net effect is that there is a shift, at the national scale, from a state of more concentration, or urbanisation, to one of less concentration, or counterurbanisation.

Despite the simplicity of the concept, the meaning of counter-urbanisation is easily misinterpreted. One source of confusion arises out of the failure to separate cause and effect. Counterurbanisation is a descriptive label which is applied to the observed pattern of change over time in the balance of urban and rural populations. No reasons for such changes are necessarily implied. The size of the population living in an area is determined by the numbers of births, deaths and migrants. Attempts to define counterurbanisation (or indeed urbani-sation) in terms of any one of these components of change are in consequence misleading: counterurbanisation is not synonymous with net out-migration.

A second difficulty concerns questions of scale. There is a clear distinction between the long-standing and continuing process of central city loss and suburb/hinterland gain, and the comparatively new phenomenon of counterurbanisation by which rural areas exhibit a resurgence of growth. The essential difference is that whereas with a shift to the suburbs and exurbs the population remains within the daily urban system, and so continues to participate in the routine life of the metropolis, with counterurbanisation there is a 'clean break' with the city. Counterurbanisation is in no way incompatible with decentralisation. Each refers to patterns of population change which may or may not be occurring at fundamentally different scales.

In the stages of development model, absolute decentralisation characterises the first phase of counterurbanisation. With the daily urban system in overall decline, it is only the ring which continues to gain in population. Later, even this area declines and as both core and ring lose population, the dominant form of change becomes that of relative decentralisation. Although counterurbanisation is now well established in many advanced economies, it represents only the most recent stage in the course of urban development. A continua-tion of the processes of deconcentration leading to more rural growth and urban decline is one possible future. Another is that of reurbanisation (Table 3). Berg *et al.* (1982) argue that urban renewal will eventually restore the appeal of the city. Although still declining overall, the daily urban system will, under these conditions, undergo a relative recentralisation as the losses in the core diminish. With a return to limited growth in the core, the centralisation process becomes absolute.

The comparative recency with which powerful deconcentration trends have become established suggests that for the foreseeable future at least, advanced Western economies will be characterised by urban population decline and counterurbanisation. Urbanisation, however, as explained in the previous section, seems to be deeply entrenched throughout much of Africa and Asia, with few signs as yet of the onset of significant exurbanisation: significant daily commuting is not possible because of low levels of private car ownership and the limited capacity of feeder road networks. The stages of development model has no predictive content, since it is purely a descriptive generalisation. Its principal value is that it provides a useful conceptual framework, based upon the experiences of Western cities, with which to generalise and to pursue explanations about long-term changes in the distribution of urban populations at both national and intraurban scales.

REASONS FOR URBANISATION

Urbanisation at the global scale is a very recent phenomenon. Although towns and cities have existed since neolithic times, a wholesale shift of population from rural to urban has occurred only in the last fifty years. With slightly over one quarter of the population living in towns and cities in 1950, the world was very much a rural place. The population was more urban than rural in North America and parts of Europe, South America and Australasia, though only in the United Kingdom and the Netherlands did more than 80 per cent live in towns and cities (Figure 14). Most of Africa and Asia, and the remaining parts of South America were rural, with fewer than 20 per cent of the population being town and city dwellers. In these areas, such urban development as existed was highly localised, and occurred predominantly in coastal pockets, so there were vast tracts of inland territory in which there was negligible urban population.

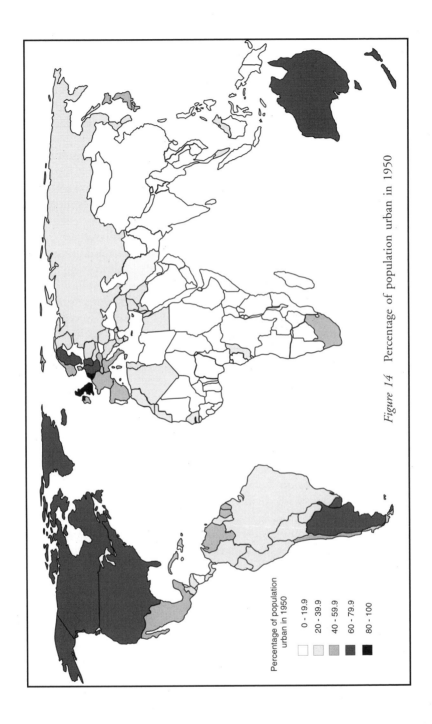

Figure 14 Percentage of population urban in 1950

Percentage of population
urban in 1950

0 - 19.9
20 - 39.9
40 - 59.9
60 - 79.9
80 - 100

The 1950 pattern bears little resemblance to that today. The differences between Figures 4 and 14 point to the operation, since mid-century, of new and powerful processes of population concentration which have extended urban patterns across the globe. As well as spatial extent, the speed of change is remarkable. It took nearly eight millennia for the population of the world to rise to 25 per cent urban: it took less than half a century to rise a further quarter to present levels. In order to understand the contemporary urban world it is therefore necessary to analyse the factors which led slowly and incrementally to the distinctive core/periphery pattern of urban development by 1950; and those responsible for rapid global urbanisation since.

Urbanisation is the consequence of processes which concentrate people in urban areas at the expense of the countryside. Although towns and cities existed in many countries in the mid-sixteenth century, overall levels of urban development were low. It is doubtful if any territory, or the world as a whole, was more than 1 per cent urban in 1550, and of the few cities, only Paris, Naples, Venice and Lyons had populations in excess of 100,000 (Chase-Dunn, 1985: p. 279). Urban development was restricted by the volume and value of surplus product that could be generated and accumulated in a single place. The low level of economic development effectively imposed a ceiling on the number of people that could be sustained in urban places.

Most researchers agree that this constraint was broken irrevocably and irreversibly when industrialisation raised levels of economic output to unprecedented levels (Clark, 1982). This formative transition led to the production of huge surpluses and the consequent emergence of the first urban society. Once self-perpetuating urbanisation had been 'invented' it was both copied and exported. The former, which may be termed indigenous urbanisation, occurred in adjacent areas where the social and economic circumstances were similar to those in the 'breakthrough' economy. It created a core region of advanced and dominant urban industrial economies. The latter, which is imposed urbanisation, took place in those dependent territories which were most closely linked with the core. Significant time-lags were involved in both processes, so that gross unevenness characterised the early stages of global urban development.

The mechanisms involved can be explained by the interdependency theory of global urban development, so called because it sees urbanisation in both core and periphery as interrelated consequences of a common set of causes. Global urban development, according to the theory, is a consequence of two linked processes; changes in the way in which wealth is accumulated, and the evolution of the world-system of nations (Table 4). Historically, the former involved a sequential development of the prevailing economic system from mercantilism, through industrial and monopoly capitalism, to transnational corporate capitalism (Castells, 1977; Goldfrank, 1979). It had its own momentum in the form of the drive for ever-higher levels of output and profit through the development of distinctive units of production and modes of accumulation. The latter was concerned with the division of the world into progressively larger spheres of economic association and exchange based upon changing space relations and systems of supply (Taylor, 1993). Each was dominated by different cities and was created under the auspices of different hegemonic powers. It culminated in the emergence of a single integrated capitalist world-economy which incorporates most nation states (Chase-Dunn, 1989). These two sets of inter-dependent structural and spatial changes set in motion circular and cumulative processes of population concentration, in a manner described by Pred (1977) and outlined in Chapter 2, that affected the world at different times and in different places over the past three centuries.

It is important to emphasise that interdependency theory proposes a single explanation or interpretation for urbanisation, whether in developed or in developing economies. Urban development in both is seen as the product of the emergence of a capitalist mode of production, and its subsequent evolution and changing space relations. They are different consequences of the same cause. This perspective has powerful antecedents in dependency theory, which explores and attempts to account for the links between development in core regions and underdevelopment in the periphery (Frank, 1967, 1969). The principal argument in this dependency school of thought is that underdevelopment in one part of the world is a corollary of development in another. It is a result of the plunder and exploitation of peripheral economies by economic and political groups in core areas. Urbanisation can be seen in a broadly similar way, but by taking the argument one stage further it is suggested that urban development, wherever it occurs, is one of the spatial outcomes of capitalism. When seen from the developing world most urbanisation appears to be 'dependent' in the sense that it is introduced or imposed by the developed world. From a global perspective, however,

Table 4 Principal stages in global urban development

	1500–1780	1780–1880	1880–1950	1950–
MODE OF ACCUMULATION				
Economic formation	Mercantilism	Industrial capitalism	Monopoly capitalism	Corporate capitalism
Source of wealth	Trade in commodities	Manufacturing	Manufacturing	Manufacturing and services
Representative unit of production	Workshop	Factory	Multinational corporation	Transnational corporation, global factory
WORLD-SYSTEM CHARACTERISTICS				
Space relations	Trade routes	Atlantic basin	International	Global
System of supply	Colonialism	Colonialism/ imperialism	State imperialism	Corporate imperialism
Hegemonic powers	United provinces, Mediterranean city states	Britain	Britain, USA	USA
URBAN CONSEQUENCES				
Level of urbanisation at start of period (per cent)	2	3	5	27
Areas of urbanisation during period	European ports	Britain	North western Europe, the Americas, coasts of empires	Africa and Asia
Dominant cities	Venice, Genoa, Amsterdam	London	London, New York	New York, London, Tokyo

all urbanisation can be held to be interdependent in that it stems centrally from capitalism and its spatial relations. This is not to say that all urbanisation has arisen in an identical way and is therefore the same in all countries. Capitalism has adopted different forms at different times, so producing spatially differentiated patterns of urban development at the global scale.

The interdependency theory of global urban development can be criticised on three principal grounds. The first is that urbanisation in the developing world lagged so far behind that in the developed world that it cannot be regarded as part of the same process. Britain was an urban industrial society for three-quarters of a century before any territory in what is now the developing world passed the 50 per cent urban threshold, and the urbanisation of the periphery did not gather real momentum until after 1950. In considering this argument it is important, however, to place urbanisation in its context of space and time. Global urbanisation involves massive shifts in the distribution of population over a wide area and is inherently a slow process. It is perhaps no accident that self-sustaining urban development first occurred in a very small country where forces of urban growth were concentrated and reinforced each other. Elsewhere, and especially in the periphery, they were dispersed over a wider area and so took longer to have an effect. The importance which is assigned to time lags is in part a function of perspective. When looking back over the last two centuries from the present, lags of a few decades appear to be of major significance. When viewed in the context of the eight millennia which have elapsed since the foundation of the first cities, they appear trivial.

A second criticism is that independency theory understates and underestimates the rich traditions of urban development, supported by economic systems that were not explicitly capitalist in formation, which existed in developing countries. Proponents of this point of view point to the highly successful urban civilisations in ancient Egypt, India, China, Cambodia, Peru, Mexico and Nigeria which produced great cities including Luxor, Delhi, Hyderabad, Cuzco, Angkor, Tenochtitlan, Anyang and Chengchow. Such urban developments, they argue, were products of states and economic systems which were religious, military or feudalistic in formation. Even allowing for the more parochial scale of spatial organisation in these early societies, the significance of capitalist exchanges in what were then world economic systems, was highly marginal.

Independency theory, however, recognises and indeed emphasises the achievements of pre-industrial economies, although it is argued that they were largely incidental to global urban development. Levels of productivity and surplus in early urban hearthlands were never high enough to facilitate self-sustaining urban development and so their importance was temporary and local and had few wider consequences. The urban civilisation of ancient Egypt, for example, was highly successful in cultural and scientific terms. Although it lasted for several millennia it failed to act as a catalyst for parallel urban development elsewhere in North Africa and Western Asia. Rather than denying and devaluing their contribution, independency theory provides a powerful explanation as to why pre-industrial urban economies were not more successful.

The final criticism is that capitalist theories do little more than state the obvious, and often in a language that serves to obscure rather than to clarify. Capitalism is the prevailing economic formation in all but a small minority of countries where it is excluded by totalitarian governments. To say that it causes urbanisation is to advance explanation and understanding very little, as all social outcomes, both structural and spatial, are the products of capitalism. Such arguments have some validity at the most general level, but they fail to distinguish between capitalism as an underlying principle and as a specific and evolving economic formation. The value of independency theory lies not in its foundations in capitalism *per se*, but in the links which it proposes between successive stages in the development of capitalism and urban growth, urbanisation, and the proliferation and increase in size of towns and cities.

STAGES OF GLOBAL URBAN DEVELOPMENT

The foundations for urban development in the core, and in localised areas in the periphery, were established up to about 1780 under conditions of mercantilism. This was an economic system which originated in the fifteenth century and involved the accumulation of wealth through trade. Its main feature was the buying and selling of the products of labour, principally agricultural and craft items. The aim of the merchant, as typified by Antonio in Shakespeare's *The Merchant of Venice*, was to buy goods at one price and sell them at a higher price, and to consume some of the profits and reinvest the remainder in further trade. The highest profits could be obtained from long-distance trade in scarce commodities, so Antonio's ships were laden with exotic silks and spices from the orient. Cloves were the most highly prized commodity in the mid-sixteenth century, being more valuable than the equivalent weight in gold! Most trade followed established land or sea routes and took place with suppliers in well defined source areas, so the Mediterranean, the Baltic, the North Atlantic, the Indian ocean, and central Asia emerged as distinctive trade areas. Buying, selling and consumption, however, were restricted to towns and cities, where sources of finance, trading opportunities and good communications were available. Any increase in the volume of trade therefore led to urban development. Under mercantilism, wealth generally originated in rural areas, whereas expenditure and consumption took place in urban centres (Johnston, 1980).

An important feature of mercantilism was the belief that the volume of trade was finite, so wealth could best be accumulated by capturing supplies and markets from rivals. This competition for territory and its products was the prime driving force behind exploration, discovery and colonisation in the sixteenth and seventeenth centuries. It was a process that was endorsed by national governments and was led by monarchs, aristocrats and privately owned companies who sponsored exploration and settlement in the hope that it would lead to new trading opportunities and the creation of vast wealth. An example is the Plymouth and Virginia company which was established in 1615 with the express purpose of settling the Chesapeake Bay area of North America so as to produce the wine, silks and spices to compete with those that were the monopoly of Spanish merchants. In the event, the climate was found to be unsuitable, and the principal crop was tobacco. The British and Dutch East India Companies, and the Royal Niger Company are similar examples of early colonial organisations which were set up to promote settlement and to develop trade.

Mercantilism was responsible for establishing the foundations for urban development in colonial powers. It made possible the introduction of highly profitable trading links which led to the generation and concentration of wealth in cities. In 1500, the world's ten largest cities were all in Europe (Table 5). At the other end of the trade route, there was only limited urban development as the role of the periphery was to supply, but not to process, basic agricultural products and raw materials. For example, the southern colonies of North America, on Chesapeake Bay, grew and shipped tobacco directly to Britain where it was cured, packed and sold. By 1800 there were

Table 5 The world's ten largest cities in descending order of size, 1550-1991

1550	1700	1900	1991
Paris	London	London	Mexico City
Naples	Paris	New York	New York
Venice	Lisbon	Paris	Los Angeles
Lyons	Amsterdam	Berlin	Cairo
Granada	Rome	Chicago	Shanghai
Seville	Madrid	Philadelphia	Beijing
Milan	Naples	Tokyo	Seoul
Lisbon	Venice	Vienna	Calcutta
London	Milan	St Petersburg	Moscow
Antwerp	Palermo	Manchester	Paris

Source: updated from Chase-Dunn (1985)

the beginnings of urban development in the Greater Caribbean (from Maryland to north east Brazil), the middle Atlantic and New England areas of North America, the East Indies, and along the coasts of Africa, India and China (Taylor, 1993).

The process of colonisation extended over many decades and so led to the creation of urban patterns of varying complexity. In some territories there were existing urban structures upon which colonial influences were introduced, but mostly there was no prior urban settlement of significance and so the pattern which developed was wholly colonial in character. The ways in which towns and cities were introduced into and grew in areas in which there was little or no existing urban development, are described and explained by Vance (1970) in his mercantile model of settlement evolution which is based upon the east coast of North America. A modified version is shown in Figure 15. The model identifies a sequence of stages through which initial contact leads to differential urban development in the core and the periphery. For Vance, the first phase of mercantilism involved the exploration of overseas territory and the search for information about the potential for production and trade. Once favourable reports were received, staple products such as fish, fur and timber were harvested, but no permanent settlement was established (stage 2). Urban development began in stage 3 with the first settlement of colonists after 1620 who both produced staples and consumed products manufactured in the home country. The two-way trading links which were established focused upon the principal port, which became the administrative centre of the colony.

Colony   **Home area**

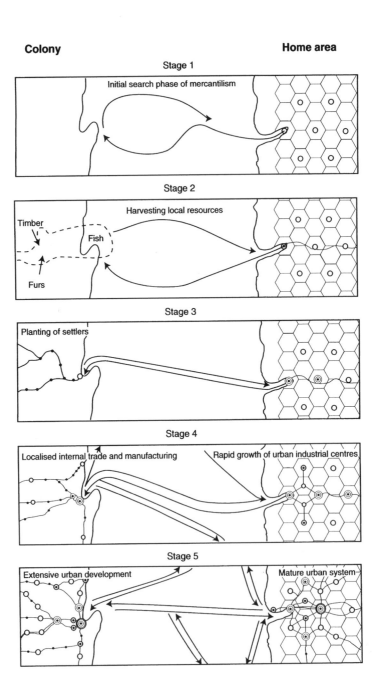

Figure 15 The mercantile model of urban development
Source: adapted from Vance (1970)

The subsequent stages trace the development of the urban system under capitalism. The fourth period is distinguished by the development of internal trade and manufacturing in the colony and the extension of transport links and trading routes from the major gateways into the interior. At the same time, there is rapid growth in manufacturing in the homeland to supply both the overseas and the domestic market. The final stage is achieved when a fully fledged mercantile settlement pattern exists in the colony and is matched by a mature industrial system of cities, organised on a central place basis, in the imperial power.

An important characteristic of the settlement pattern in the colony is its linearity. Settlements are aligned along the coasts and also are located along the routes of trade which connect the coastal points of attachment to the staple-producing interiors. Over time these become integrated within a functionally interdependent system of cities (Friedmann, 1972). Close historical and geographical parallels exist between the patterns of urban development in the United States, as modelled by Vance, and those in West Africa and Brazil, as identified by Taaffe, Morrill and Gould (1963). The work undertaken by these authors suggests that the principal contribution of mercantilism was the creation of a well developed urban pattern in the core, but an embryonic urban framework, except at points of contact where exporting centres flourished, in the periphery.

URBANISATION IN A CAPITALIST WORLD-ECONOMY

The spatial patterns of urban development which were established under mercantilism and early colonisation were accentuated and compounded when capitalism and imperialism became the dominant economic and political systems in the late eighteenth century. Capitalism is a form of economic organisation in which wealth is generated for investors through the production of saleable goods and services. Its main feature is that the capitalist employs workers directly, rather than merely bargaining for and trading in the items which they produce. Profits are made from the differences between the value of the products of labour, and the price which is paid for it. To be most successful it requires large inputs of raw materials, and extensive markets, which are best ensured through the possession of empire. Capitalism, through mass production and associated agglomeration, generates urban growth and urbanisation because it concentrates productive activity and all the workers and spending power that are associated with it. In addition, the city continues to serve as a centre for the consumption of the profits of capitalism. Under mercantilism, cities are the points of consumption and the articulation of trade. Under capitalism there is a third function, organised mass production, which is linked to the other two (Johnston, 1980). At the same time, the increased volume of trade stimulates additional urbanisation in dependent overseas territories.

With the benefit of hindsight it is possible to trace the evolution of capitalism through distinct industrial, monopoly and transnational corporate stages since it first became the dominant economic system in late eighteenth-century Britain (Table 4). The individual stages are distinguished by both structure and space relations. The former relates to the underlying economic formation and is reflected in the representative type and unit of production. The latter refers to the geographical locations and areas in which wealth accumulation principally occurred. Each stage gave rise to, and accelerated, urban development in different parts of the core and the periphery.

The initial phase was industrial capitalism, in which wealth was created by making rather than merely trading in goods. Manufacturing involved large numbers of people who tended machines which performed sets of routine and repetitive operations in order to make standard products. Large inputs of raw materials were involved, so industrial capitalism was supported through extensive trade with overseas suppliers of fibres, ores and agricultural products. Many were controlled through colonial administration. Although small by today's standards, the early industrial factory employed many more people than the craft workshop which it succeeded. As such it generated sizeable concentrations of population and labour, which in turn attracted more industry and so led to further urban growth. Increased demand was similarly translated into urban development in economically linked parts of the periphery.

Great Britain was the first country to enter into and progress through the phase of industrial capitalism. The industrial revolution, which began in the last third of the eighteenth century, transformed the country from a rural agricultural economy to an urban industrial economy in less than one hundred years. The pace of population growth was unprecedented and unparalleled. At the first census in 1801, the total population of England and Wales was some 8.9 million. By 1891, the last census of the century, it had risen to 29.0

million, an increase of 326 per cent. The rate of urban growth, however, was even more impressive as the urban population grew by 946 per cent (Carter and Lewis, 1991: p. 32). Between 1801 and 1851 alone, over nine million people were added to the population of England and Wales, but while those living in places with fewer than 5,000 people increased from 6.6 to 9.9 millions, the town dwellers increased from 2.3 to 8.0 millions (Weber, 1899). Of the total increase, 64 per cent fell to the towns and cities. A population that was 26 per cent urban in 1801 was 45 per cent urban in 1851. By 1861, for the first time in any country, more people in England and Wales lived in towns and cities than lived in rural areas.

Industrial capitalism created a new pattern of urban settlement in Britain termed by one contemporary observer 'the age of great cities'. At the beginning of the nineteenth century, London, with some 861,000 people was the largest city in the world, exceeding Constantinople (570,000) and Paris (547,000) but it was the only place in Great Britain with over 100,000 people. By 1851 its population had risen to 2.4 million and there were two other British cities, Liverpool and Manchester, with over 300,000. Birmingham, Leeds, Bristol, Sheffield and Bradford had between 100,000 and 300,000 and there were a further 53 cities between 10,000 and 100,000 in size.

In contrast to its profound impact upon settlement patterns in Great Britain, industrial capitalism did little to change the urban/ rural balance of population elsewhere. Although by 1890, industrialisation had spread to adjacent parts of north western Europe and to North America, the level of urban development in the core remained low. It was negligible, when measured at the national scale, in the periphery. It is this urban world which was analysed and illustrated in detail by Weber in his classic work on *The Growth of Cities in the Nineteenth Century* (1899). The limited extent of urbanisation at the time is apparent when his data are mapped as far as is possible according to the present network of state boundaries. It goes without saying that this is a task of very crude approximation which can only be undertaken with considerable cartographic licence, because the political geography of the world has changed so much over the past century (Figure 16). Only three areas – Great Britain, northwest Europe and the USA – were more than 25 per cent urban in 1890. With less than 3 per cent of the world's population living in towns and cities, there was little or no urban development in most other territories.

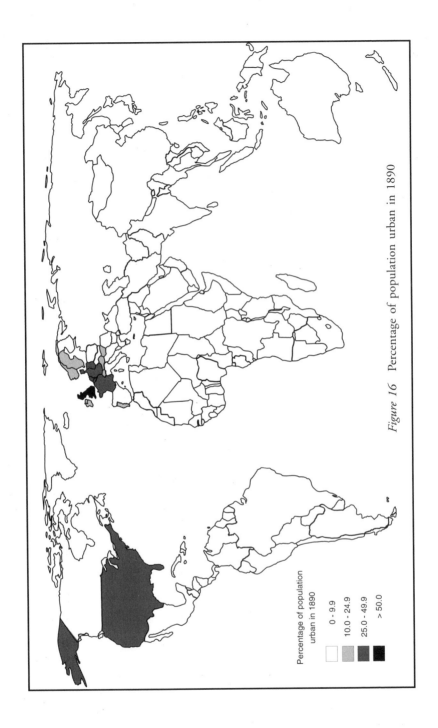

Percentage of population
urban in 1890

0 - 9.9
10.0 - 24.9
25.0 - 49.9
> 50.0

Figure 16 Percentage of population urban in 1890

Such localised urbanisation as was produced in the periphery was, in accordance with Vance's model, a consequence of the concentration of population around points of supply (Figure 15). The industries of the core used domestic coal and iron ore to build and power machines to process cotton, sugar, jute, rubber, tobacco, wheat, tea and rice imported from colonies and imperial territories. These commodities were the products of agriculture, and their primary accumulation in the periphery led to limited urban development, associated with shipment rather than with local processing. For example, cotton was produced on plantations in the USA, from where it was exported for spinning and weaving in Europe. This trade gave rise to urban development in New England, where the shipping companies and financial interests were based, and in ports in the south, especially Charleston and Savannah, but the interior plantation south remained rural. By 1870, about 30 per cent of the American population was urban, although all the main towns and cities were in the north east and Great Lakes regions (Borchert, 1967). Elsewhere in the periphery, the growth of cities was linked to export production and international trade. 'São Paulo grew on the basis of coffee, Accra on cocoa, Calcutta on jute, cotton and textiles, and Buenos Aires on mutton, wool and cereals' (Gilbert and Gugler, 1992: p. 47). Urban development in association with peripheral supply similarly took place in the West Indies and Indonesia, Malaysia and the Far East. Although cities were established along the coasts of empire, these developments did little to change the overwhelmingly rural distribution of the local population.

The reason for the limited impact was that the towns and cities which were established under colonialism were more closely linked to the urban system of the European power than they were to settlements in the surrounding area. The purpose was to facilitate economic imperialism rather than to service or promote economic development in the colony. The process of urban formation within the British empire is documented in detail by King (1990). In 1800, the principal colonial cities were Calcutta, Bombay, Madras, Dhaka, Sydney, Halifax, Montreal, Toronto, Port of Spain, Bridgetown, Gibraltar, Kingston and Nassau. A century later, by 1900, this list had expanded to include Aden, Hong Kong, Cape Town, East London, Durban, Pretoria, Johannesburg, Salisbury, Blantyre, Mombasa, Kampala, Zanzibar, Lagos, Accra, Nicosia, Suez, Port Louis (Mauritius), Mahé, Kuching, Georgetown (Guiana), Melbourne, Brisbane, Adelaide, Perth, Hobart, Christchurch, Wellington, Port

Table 6 Principal characteristics of colonial cities

Geopolitical
1. External origins and orientation

Functional
2. Centre of colonial administration
3. Presence of banks, agency houses and insurance companies
4. Focus of communications network
5. Warehousing/distribution centre

Economic
6. Dual economy, dominated by foreigners
7. Presence of large numbers of indigenous migrant workers
8. Municipal spending biased towards colonial elite
9. Dominance of tertiary sector
10. Parasitic relations with indigenous rural sector

Political
11. Eventual formation of indigenous bureaucratic-nationalist elite
12. Indirect rule through leaders of various communities
13. Social polarity between superordinate expatriates and subordinate indigenes
14. Caste-like nature of urban society
15. Heterogeneous society comprising colonial elite, in-migrants from other colonial territories, in-migrant educated indigenes, and uneducated indigenes
16. Occupational stratification by ethnic groups
17. Pluralistic institutional structure
18. Residential segregation by race

Physical/spatial
19. Coastal or riverine site
20. Establishment at site of existing settlement
21. Gridiron street plan
22. Presence of elements of Western urban design
23. Residential segregation between colonials and indigenes
24. Large differences in population density between residential areas of colonials and indigenes
25. Tripartate division between indigenous, colonial and military cities

Source: modified from King (1990: pp. 17–19).

Moresby and Port Stanley. An important feature of these cities, emphasising their imperial connections, is that their built and spatial environments had much in common with each other but were different to urban centres in the interiors of the countries in which they were situated (King, 1990: p. 140). Colonial cities had a very distinctive role as nerve centres of colonial exploitation, for they were places where the banks, agency houses, trading companies and shipping lines, through which capitalism extended its control over the local economy, were situated (McGee, 1967). A massive literature exists on the functions and features of the colonial city. Telkamp (1978), in a comprehensive overview reproduced by King (1990), identifies some thirty distinguishing characteristics (Table 6). The colonial cities were the early peripheral links in the emerging world-economy. Because of their early lead in urban growth, many subsequently became primate centres, and in some cases the capitals of independent states.

Monopoly capitalism replaced industrial capitalism and colonialism towards the end of the nineteenth century. It was distinguished by a vastly increased scale of economic activity, and the domination of newly created international markets, within state-controlled empires, by a small number of producers in each sector. Monopoly capitalism emerged in response to the demand for products that was generated by the rapidly growing population of the industrial nations. This stimulated manufacturers to diversify from making heavy, crude products into the mass production of a wide range of consumer goods and services. Increased output occurred both because the core economies in Europe became more productive, and because the manufacturing belt of the USA attained core status alongside Britain, France, Germany and the Low Countries during the 1880s (Chase-Dunn, 1989). It was achieved through the consolidation of many factory enterprises into multinational corporations which typically engaged in many functions in many areas, both at home and in the periphery.

Monopoly capitalism involved the more ruthless exploitation of peripheral areas. The larger scale of industrial activity required the international sourcing of raw materials and the international marketing of manufactured products, so the success of the core regions became dependent on their ability to dominate and control overseas territories. This was either through formal imperialism, of which the British and the French empires were the largest manifestations in the early twentieth century, or else through corporate power and influence as increasingly exercised, for example, by the industries of the USA. Britain established itself as the leading imperial power after about 1880 when it increasingly drew its industrial raw materials, including ores, oil and rubber, from around the world and in return supplied its overseas possessions in India, Africa and the Far East, and other territories, with railways, ships, machinery, arms and motor vehicles. Similarly, the USA rapidly became a major international player after 1909 when, symbolically, Selfridge's store was opened by an American-born British merchant in Oxford Street, London, at the very centre of the dominant power in the world-economy (King 1990; p. 81). Thereafter, many of the major US corporations, including Goodyear, Standard Oil, Ford and General Motors, developed international spheres of operation.

Monopoly capitalism produced further urban growth and urbanisation in an expanded core, although urban development in the periphery remained limited. Precise comparison of the urban world in 1890 (Figure 16) with that in 1950 (Figure 14) is inappropriate because Weber's data for the nineteenth century are far less reliable and refer to a very different geopolitical era. The overall pattern of change, however, is clear. Urbanisation in the first half of the twentieth century occurred most rapidly and extensively in Europe, the Americas and Australasia. Most of the rest of the world was unaffected.

Much of the urban pattern in 1950 is explained by processes of population concentration that were associated with the economic and political imperialism of the United States, Russia, the United Kingdom and France. High levels of urban development in Canada, South and Central America were a legacy of British trade and, more recently, of corporate links with the USA. Limited urban development existed across the Russian empire in Asia, central and eastern Europe. Urbanisation elsewhere in the periphery was largely a product of British and French imperialism. Although only one-quarter of the population lived in urban places, the principal feature of the urban world in 1950 was that the cycle of urbanisation in the dominant economies of the core was, or was very nearly, ·complete. In the periphery it had hardly begun.

CONCLUSION

The focus of this chapter has been placed upon the historical processes of urban growth and urbanisation at the global scale. Emphasis has concentrated on the forces which were responsible for the creation of large cities and societies with predominantly urban populations in the developed and the developing worlds. Both urban growth and urbanisation have long histories and were accelerated under mercantilism, but the progression towards the contemporary urban world did not gain any significant and sustained momentum until the industrialisation, in the early nineteenth century, of some of what are now numbered among the core economies. This key development was a consequence of the emergence of industrial capitalism as a dominant economic and social formation and of its external relationships which were formalised through colonialism and imperialism. It led to major and rapid urban development in the core economies of north western Europe and North America and subsequently, through relationships of interdependency, established the foundatiohs for urban development in selected locations in the periphery.

Urban growth and urbanisation were reinforced and extended by monopoly capitalism in the late nineteenth century. The urban development which resulted was largely restricted to the core areas and to the coasts of empire, so that the world in 1950 was highly differentiated in urban terms. The next chapter examines the progressive emergence of an urban world over the past fifty years as urban growth and urbanisation spread, under corporate capitalism, throughout the periphery.

RECOMMENDED READING

Breeze, G. (1972) *The City in Newly Developing Countries*, Englewood Cliffs NJ: Prentice Hall.

A set of readings on urbanism and urbanisation in the first half of the twentieth century. It includes the seminal paper by Kingsley Davis on the urbanisation of the human population.

Clark, D. (1989) *Urban Decline*, London: Routledge.

An analysis of patterns of counterurbanisation and urban decline using evidence and examples from Britain.

Johnston, R. J. (1980) *City and Society: An Outline for Urban Geography*, Harmondsworth: Penguin.

An excellent explanation and critique of urban growth and urbanisation under mercantilism and different phases of capitalism.

King, A. D. (1990) *Urbanism, Colonialism and the World-Economy. Cultural and Spatial Foundations of the World Economic System*, London: Routledge.

A detailed and comprehensive overview of the literature on the contribution of colonialism to urbanisation at the global scale. The book explores the links between the metropolitan core and the colonial periphery and assesses the contribution of colonialism in creating the urban characteristics of the two areas.

Gilbert, A. and Gugler, J. (1992) *Cities, Poverty and Development*, Oxford: Oxford University Press.

A comprehensive analysis of the causes, characteristics and consequences of urbanisation in Africa, Asia and Latin America. The chapter by Gilbert on urban development in a world system is especially useful.

Potter, R. B. (1985) *Urbanisation and Planning in the Third World*, Beckenham: Croom Helm.

The first two chapters provide a useful general overview of theories of urban growth and urbanisation with an emphasis upon the Third World. Later chapters explore trends and problems in the Caribbean.

Taylor, P. J. (1993) *Political Geography: World-Economy, Nation-State and Locality*, London: Longman.

A general political geography of the world-economy which includes a highly detailed account and explanation of the rise and fall of colonies and world empires.

4

URBAN DEVELOPMENT AS A GLOBAL PHENOMENON

The world has recently become an urban place principally because of major changes in the distribution of population in developing countries. Until mid-century, urbanisation was a process that was largely restricted to the core regions of the mercantilist, industrial capitalist and monopoly capitalist economic systems, where it produced high levels of urban development across large parts of Europe and North America. In the fifty years since, urban growth and urbanisation have affected the rest of the world. Behind this shift lie major changes in global patterns of accumulation and consumption which have profoundly different urban geographical consequences. On the one hand they have concentrated global economic power in a small number of world cities in the core economies, as is discussed in Chapter 7. On the other they have extended and accelerated urban development throughout the periphery.

Urbanisation became a global phenomenon as a consequence of far-reaching changes in the structure and spatial relations of capitalism. Two principal developments were involved: the replacement of monopoly capitalism by transnational corporate capitalism, and the creation of patterns of production, trade and service provision which, rather than being restricted to the north Atlantic, or to political empires, are truly global in extent. The former represents the emergence of a mode of production which is suited to accumulation in a rapidly expanding world market. The latter is a response to the ending of the opportunities for further imperialism, and the achievement of political independence by many colonies (Becker *et al.*, 1987). The development of transnational corporate capitalism and global exchange do not, however, represent radical departures from what went before. Rather, they are the most recent stage in

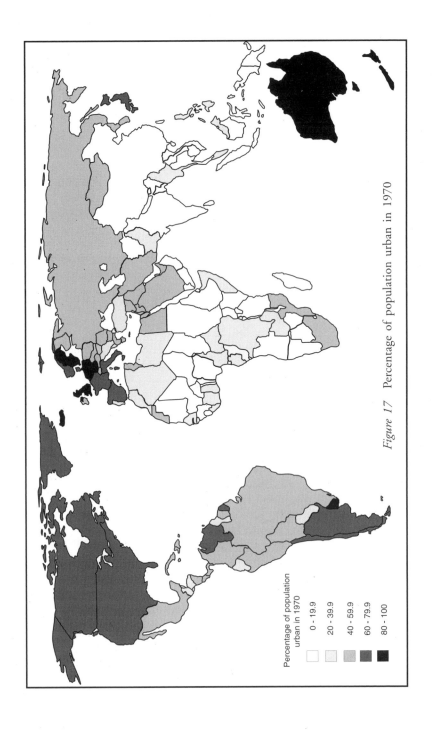

Figure 17 Percentage of population urban in 1970

Percentage of population
urban in 1970

0 - 19.9
20 - 39.9
40 - 59.9
60 - 79.9
80 - 100

the evolution of capitalism (as shown in Table 4) together with its space relations, as a system of wealth accumulation (Chase-Dunn, 1989). Together these developments define a new world-economic order with distinctive structural and spatial characteristics, a principal consequence of which was and is rapid urbanisation in many of the world's peripheral areas.

It is important to emphasise the speed with which urban development has transformed patterns of settlement across the world over the post-war period. The changes are captured by maps which show levels of urbanisation in 1950, 1970 and 1990 (Figures 14, 17 and 18). Africa and Asia were almost wholly rural in 1950 and it is here that the subsequent transition to urban living is most marked and has had, and is having, the most profound consequences. Significant urban development, measurable at the national scale, began to affect parts of Africa and the Middle East between 1950 and 1970, although many countries in these regions, especially in the Sahel belt of Africa and in south west Africa, were largely unaffected. No country in Africa was more than 50 per cent urban in 1970.

Urbanisation began significantly to affect the countries of South and East Asia somewhat later, as most did not pass even the 20 per cent urban mark until 1990. This lag, involving the highly populated countries of China and India, as well as Indonesia, Bangladesh and Pakistan, meant that the level of urbanisation at the global scale remained low (Figure 12). Levels of urbanisation across the periphery were higher in 1990 than in 1950, although the pattern in Africa and Asia was highly varied. Average levels were well below those in South America and in the core. There remained a small number of countries in the remoter regions of Africa and Asia in 1990 with fewer than 20 per cent of their population living in urban places (Figure 18).

THE NEW ECONOMIC ORDER

The extension of urban development as a global phenomenon is widely seen as a consequence of the reorganisation of production, labour, finance, service provision and competition, on an international, and in some cases a transnational basis. Over the past half century, an increasing share of production has been organised globally, rather than within the narrow confines of nation-states or empires. Much production has shifted to the periphery both as a means of penetrating local markets and so as to use cheap labour

79

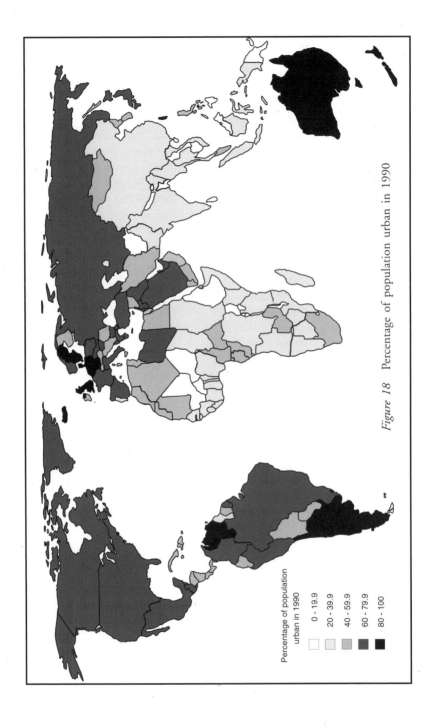

Figure 18 Percentage of population urban in 1990

Percentage of population
urban in 1990

0 - 19.9
20 - 39.9
40 - 59.9
60 - 79.9
80 - 100

to make goods for sale in the core economies and elsewhere (Frobel, Heinrichs and Kreye, 1980). Examples of industries where this has happened include electronic goods, drugs, motor vehicles, clothing, machine tools and domestic appliances. At the same time, several countries in the developing world have expanded their manufacturing capabilities and the firms in these newly industrialised economies have captured markets for their products in the developed world. The production of some foodstuffs has also been reorganised on a commercial basis so that it can be exchanged globally. Domestic agricultural production in many developing countries has been replaced by production for export, a beneficial consequence, as far as global capitalists are concerned, being that it generates currency which can be used to purchase more imports and so increase external dependency.

The transnationalisation of production involves the manufacture of global products, with global brand names, which are assembled across the world from components made in a number of countries. It is achieved by direct investment by firms from the core economies in developing countries, a practice that increased significantly in the 1980s (Sassen, 1994). Firms which organise production in this way are able to take advantage of local conditions, especially the availability of large pools of very cheap labour, so as to reduce manufacturing costs and maximise world-wide profits. Ford's 'world car' which is designed, produced and built in strategically located plants from parts fabricated in both developed and developing economies is an example of a product which is made in this way. The semiconductor industry is similarly organised on a global basis (Castells, 1984). Such are the numbers of people who are incorporated within international systems of manufacturing that it is appropriate to talk of the emergence of a 'global factory' as the representative unit of production under transnational corporate capitalism.

Global production is principally undertaken by companies that make many products in many places. They have been responsible for a major expansion in world trade over the past twenty-five years. In the 1980s, transnational corporations were involved in between 70 and 80 per cent of world trade outside the centrally planned economies (Feagin and Smith, 1987: p. 3) They were responsible for as much as one-quarter of total world production in market economies (Dicken, 1992: p. 48). The United Nations Centre on Transnational Corporations identified some 36,600 transnational corporations in 1991, together with more than 170,000 affiliated

Table 7 The largest non-financial transnational corporations, 1990

Rank	Corporation	Country	Industry	Foreign assets (US$ bn.)	Total assets (US$ bn.)	Foreign employment	Total employment
1	Royal Dutch Shell	United Kingdom/ Netherlands	Petroleum refining	69.2	106.4	99,000	137,000
2	Ford	United States	Motor vehicles and parts	55.2	173.7	188,904	370,383
3	GM	United States	Motor vehicles and parts	52.6	180.2	251,130	767,200
4	Exxon	United States	Petroleum refining	51.6	87.7	65,000	104,000
5	IBM	United States	Computers	45.7	87.6	167,868	373,816
6	British Petroleum	United Kingdom	Petroleum refining	31.6	59.3	87,200	118,050
7	Asea Brown Boveri	Switzerland	Industrial and farm equipment	26.9	30.2	200,177	215,154
8	Nestlé	Switzerland	Food	N/A	28.0	192,070	199,021
9	Philips Electronics	Netherlands	Electronics	23.3	30.6	217,149	272,800
10	Mobil	United States	Petroleum refining	22.3	41.7	27,593	67,300
11	Unilever	United Kingdom/ Netherlands	Food	N/A	24.7	261,000	304,000
12	Matsushita Electric	Japan	Electronics	N/A	62.0	67,000	210,848
13	Fiat	Italy	Motor vehicles and parts	19.5	66.3	66,712	303,238
14	Siemens	Germany	Electronics	N/A	43.1	143,000	373,000
15	Sony	Japan	Electronics	N/A	32.6	62,100	112,900
16	Volkswagen	Germany	Motor vehicles and parts	N/A	42.0	95,934	268,744
17	Elf Aquitaine	France	Petroleum refining	17.0	42.6	33,957	90,000
18	Mitsubishi	Japan	Trading	16.7	73.8	–	32,417
19	GE	United States	Electronics	16.5	153.9	62,580	298,000
20	Du Pont	United States	Chemicals	16.0	38.9	36,400	124,900

Source: Fortune (1991)

companies across the world (UNCTC, 1993). Transnational corporations are especially prominent in the petroleum, automobile, electronics, food, drugs and chemical sectors.

Royal Dutch Shell is reckoned to be the largest non-financial transnational in terms of foreign assets, though it is relatively small in employment terms (Table 7). General Motors, which employs some 767,000, has by far the largest workforce and the largest number of foreign workers. Asea Brown Boveri, Nestlé and Philips Electronics are the most transnational in terms of the percentage of foreign sales. The global importance of transnational corporations is underlined by the fact that in 1990, the top ten had 49 per cent of their assets and 61 per cent of their sales abroad. All of the twenty leading non-financial transnationals are based in the core group of advanced Western economies, with seven originating in the USA. Transnationals accounted for 80 per cent of international trade in the USA in the late 1980s. More than one-third of this trade was between geographically separate units within the same company (UNCTD, 1991).

Changes in the tasks undertaken by labour occurred alongside the changes in the organisation of corporate activity (Cohen, 1981). A new pattern of specialisation emerged which owed less to traditional distinctions between core and periphery and more to the jobs which workers perform within transnational corporate empires. The basis of the new international division of labour is the direct employment of large numbers of workers in low-cost overseas territories to perform standard production tasks. Rather than peripheral supply and core area processing, which was the pattern under industrial and monopoly capitalism, the new economic order is one of peripheral production and manufacturing, and core area research, development, design, administration and control. Low-cost workers in developing countries undertake labour-intensive tasks of manufacture and assembly under the direction of, and to specifications drawn up by, technicians and managers based in the developed world.

Such organisational arrangements can be seen clearly in the semiconductor industry (Castells, 1984). Research and management functions, which require intellectual labour, are located in cities in core areas where the quality of life is high and where the presence of large research universities provides links with theoretical science and a pool of highly qualified graduates. Examples include Boston (Massachusetts), San Francisco, Los Angeles, Dallas and Phoenix, which are leading centres of university-based research and development and the headquarters of American transnationals in the computing and electronics sectors (Clark, 1984). The other processes, including mask-making, wafer fabrication, assembly and testing are dispersed in low-cost areas throughout the periphery. Labour-intensive assembly of computers and electronics products is especially important in Hong Kong, Singapore, Malaysia, Taiwan and Korea. A second example is provided by international publishing. The various tasks involved in producing a book tend now to be separated geographically, with editing and proof-reading, which are the most skilled operations, taking place in the core economies, while labour-intensive typesetting and printing is devolved to the developing world.

The internationalisation of production was made possible by and in turn gave rise to a new pattern of international finance. A global system of supply and circulation emerged in recent years in place of the bilateral funding arrangements, tied to trading blocs and dominated by governments, that existed at mid-century. The new system, like the old, is directed and controlled by the economies of the core through a small number of world cities (see Chapter 7), and helps to sustain their dominance. It differs substantially, however, in terms of the volume and nature of capital flows which are involved, and in the activities and places into which investment is directed.

The most important features of the global financial system are its size and composition. Data on the overall growth of global financial flows are lacking, but the International Monetary Fund estimates that the net flow of financial resources from industrialised to developing countries increased ninefold between 1960 and 1982 (Sit, 1993). The largest single element (around 33 per cent) consists of official development assistance in the form of grants and aid to promote economic development and welfare. Such funds are used for physical and social improvements in both urban and rural areas in developing countries. Around 15 per cent, however, is in the form of direct investment by foreign firms in economic projects and initiatives. The principal sources are companies in the United States, the United Kingdom, Japan, Germany, Canada and the Netherlands, which together contribute about 80 per cent of foreign direct investment funds. In the 1960s, foreign direct investment was concentrated in the exploitation of natural resources, particularly minerals and hydrocarbons which were exported and processed elsewhere. In the past decade there was a pronounced switch into manufacturing as transnational corporations established production facilities in the periphery. The pattern of destinations also changed with a shift away

A GLOBAL PHENOMENON

Table 8 The fifteen largest banks, 1995

Bank	HQ location	Capital (US$m.)	Business overseas (per cent)
Sumitomo	Osaka	22,120	32
Sanwa	Osaka	19,577	24
Fuji	Tokyo	19,388	33
Dai-Ichi-Kangyo	Tokyo	19,360	33
Sakura	Tokyo	18,549	–
Mitsubishi	Tokyo	17,651	21
Industrial and Commercial Bank of China	Beijing	16,782	–
Credit Agricole	Paris	14,718	–
HSBC Holdings	London	14,611	61
Citicorp	New York	13,625	49
Industrial Bank of Japan	Tokyo	13,596	23
Union Bank of Switzerland	Zurich	13,246	65
Bank America	San Francisco	12,058	–
Deutsche Bank	Frankfurt	11,723	37
Internationale Nederland Group	Amsterdam	11,068	33

– Less than 18 per cent

Source: The Banker July 1994, February 1995

from primary producing areas towards countries with emerging manufacturing economies. After 1986, foreign direct investment was channelled first into newly industrialising countries in Asia and then towards Association of South East Asian Nations (ASEAN) countries and China. Subsequent waves of significant foreign direct investment extended to Sri Lanka, Turkey, Chile and Mexico, so incorporating these countries within the global manufacturing economy (Sit, 1993).

The global financial system is mediated by a number of institutions, including the International Monetary Fund, which set policies and regulate credit and exchange rates, and the World Bank, which arranges and provides multilateral aid for development. It is facilitated by the development of multinational organisations such as the General Agreement on Tariffs and Trade, the Organisation for Economic Cooperation and Development and the North American Free Trade Agreement that seek to promote and influence trade. The system is dominated by a small number of powerful banks which rank alongside transnational corporations as global institutions (Table 8). Fourteen of the largest fifteen banks are based in the core economies, and seven are Japanese. The major banks of Europe and North America, however, handle more foreign business. It is made possible by growth of the US dollar and Eurodollars as international currencies and media of exchange. The global financial network operates through banking and capital markets which work on a twenty-four hour basis trading stocks, shares, futures and commodities. An important consequence is the growth of global investment practice through which capital, often in the form of aid, is spread widely through the periphery.

Developments in production and finance are associated with, and are in part dependent upon, the growth of the international service economy. Service activities which were once domestically bound have reorganised on an international basis so as to serve the needs of businesses operating across the globe. This trend is reflected in the rise of the advanced producer services sector which provides support services to industry. It includes insurance, accountancy, real estate, legal, advertising, research and development, public relations, and management consultancy firms. The change to global operation has been most marked in those service sectors in which the level of international activity was historically limited. One such field is accountancy, where the principal firms, based upon the number of audits of the largest 500 companies as identified in *Fortune* magazine, are sizeable multinationals (Thrift, 1987). The same applies to estate agencies. Global business is further facilitated by means of the organisation of employee services, including hotel accommodation, car hire and personal finance, on an international basis.

The new economic order emerged alongside, as part-cause and part-consequence of, a new political geography. By far the most important feature was the ending of imperialism by Britain, France, Belgium and the Netherlands and the attainment of political independence by many colonial territories in Africa and Asia between 1950 and 1980 (Figure 19). This added further changes to the political map which had been transformed during the 1940s by the post-war redrawing of boundaries in Europe and by the withdrawal of the British from the Indian sub-continent. Together these developments produced a large number of new nation-states which were keen to participate in the world economy so as to enjoy the benefits of trade and aid.

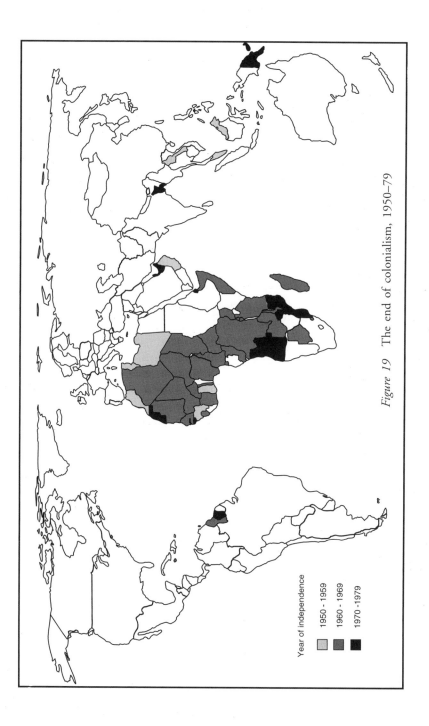

Year of independence

1950 - 1959

1960 - 1969

1970 - 1979

The new pattern was created in conditions of relative peace and prosperity, certainly in comparison with those which prevailed in the previous half century with its two world wars and numerous regional conflicts. Global stability since 1945 principally arose out of the balance of power between the West and the communist bloc, under which major wars were restricted to those in Korea and Vietnam. Localised disturbances and civil wars, often associated with decolonisation, were common among the newly independent states of Africa and Asia but although some were protracted, few of these conflicts escalated beyond national boundaries. The most important reasons were the political influence of the superpowers, especially the United States, and the mediating role of the United Nations.

Stability was further facilitated by the creation of supranational and international organisations by many of the market economies, to undertake some of the traditional roles of the nation-state. The main concerns were with defence and economic and social policy. Examples include the defensive alliance of nations in the North Atlantic Treaty Organisation and the common market of economies in the European Union. The effect was to raise overall levels of international confidence and so create improved conditions for a restructuring of capital, for purposes of wealth accumulation, on a global basis.

GLOBAL URBAN CONSEQUENCES

The new economic order is principally responsible for the recent rapid urbanisation of the periphery which in turn raised the level of urbanisation at the global scale beyond the 50 per cent mark. Since mid-century, and especially over the past twenty years, the global economy has penetrated local and regional economies across the world so that most of the remaining peripheral countries and territories have been drawn into the world-economic system. Their role, within the expanded world-economy, is through cheap labour, to produce low-cost manufactured goods and to act as markets for the products of the developed world. The effect was to raise levels of urbanisation in South America, where there were already sizeable urban populations, and to initiate urbanisation across large parts of Africa and Asia where, in 1950, it had barely started.

Transnational corporate capitalism produced and is producing urbanisation in the periphery both directly, as a consequence of urban growth in response to localised investment, and indirectly through its impact on traditional patterns of production and employment.

The former arises because economic exchanges between core and periphery are spatially focused and so lead to a concentration of globally related economic activity in urban places. Cities, especially national capitals and those with major ports or international airports, offer overwhelming advantages for profitable investment, affording wide access to cheap labour and to domestic markets. Such places are typically the major and in some cases the only centres in a country to have large-scale industry, hospitals, universities, media services and facilities for sport and the arts. As cosmopolitan centres with good external connections they are attractive to corporate managers and specialist workers on overseas postings. They are likely to be the home base of local elites which shape behaviour and consumption patterns towards which others in the country aspire.

The urban concentration of foreign investment-led economic activity is high across much of the periphery. In Indonesia, Forbes and Thrift (1987) found that overseas investment was largely restricted to the area around Jakarta, where all major foreign corporations had their headquarters. Abidjan, the capital of the Ivory Coast, has 15 per cent of the national population but accounts for more than 70 per cent of all economic and commercial transactions in the country. Bangkok accounts for 86 per cent of gross national product in banking, insurance and real estate, and 74 per cent of manufacturing, but has only 13 per cent of Thailand's population. Lagos, with 5 per cent of Nigeria's population, accounts for 57 per cent of total value added in manufacturing and has 40 per cent of the nation's highly skilled labour (Kasarda and Parnell, 1993: p. ix).

Metropolitan concentration of foreign investment is also noted in Latin America. São Paulo, with about 10 per cent of Brazil's population, contributes over 40 per cent of industrial value added and a quarter of net national product. It contains the management and production facilities of large transnational corporations and functions as a mediating point through which the domestic economy is integrated within the international market. Santiago has 56 per cent of Chile's manufacturing employment and contributes 38 per cent of national industrial output. This concentration of manufacturing activity in countries where there are few alternative sources of employment to subsistence or semi-subsistence agriculture acts as a powerful attraction to immigrants and so urban growth which is stimulated by investment is compounded by rapid population increase. For example, Sit (1993) argues that the accelerated urbanisation that stems from such investment and concentration led directly to urban growth rates which were double that of the population growth rate in most Latin American countries during the 1980s.

Urban growth and urbanisation occur in other parts of the periphery by extension. The rate of change depends upon the degree of functional and spatial integration of the domestic urban hierarchy. In countries with few other urban centres, or where they are only weakly interconnected, the extent of diffusion is limited and so the international contact points grow at a disproportionate rate. This appears to be the case in Asia, where growing international investment in manufacturing is reinforcing the position of large primate centres (Fuchs, Jones and Pernia, 1987). In others, inter-urban links are stronger and so urbanisation is, or is becoming, a nationwide phenomenon. In contrast to the historical pattern in the core, the direct urbanisation of the periphery is being imposed from the outside, rather than being generated from within. In this respect it differs only in scope and scale from the processes of peripheral urbanisation that are associated with preceding forms of mercantile, industrial and monopoly capitalism.

Urbanisation is also taking place as an indirect consequence of the impact of transnational corporate capitalism upon the economies of peripheral countries. The central argument here is that major structural adjustments are forced upon peripheral economies as the price, or penalty, for incorporation within the world-economy, and that these lead to the release of large numbers of workers from traditional occupations, who flock into the towns and cities and so contribute to urban growth and urbanisation. Peasant farmers are foremost amongst those in developing countries whose livelihoods are undermined by the drive for production of goods which will generate foreign currency, both to help reduce national indebtedness and to enable governments to acquire the symbols of statehood, such as grand presidential palaces and national airlines. Many peasant farmers have been displaced from their traditional lands and means of subsistence by the introduction of commercial agriculture which is geared to the production of exotic fruits and out-of-season vegetables for developed world consumers. They include large numbers of the very poor who have no alternative sources of employment and who must look to the city for survival. Droughts and civil wars, especially in parts of Africa, have further undermined the viability of traditional farming, leading to increased rural-to-urban migration.

The policies of post-colonial governments have stimulated urban growth by further enhancing the attractiveness of towns and cities at the expense of rural areas. One way is through the exaggerated bias of government expenditures on infrastructure and services in favour of urban areas and against the rural sector. Another is the higher wage rates and better employment protection that exist in cities because urban workers are organised into trade unions. A third is the effect of trade tariffs on the price of goods which discriminate most against low-income peasant consumers, while a fourth is the decline in the demand for locally produced staples as urban consumers develop a taste for imported food items. Such policies are creating 'backwash urbanisation' by destroying the vigour of rural areas and suffocating the cities with the burden of human casualties this process creates. The implications are seen in the rapid growth and dire social and environmental conditions of many African cities, and others throughout the developing world which are swamped by large numbers of in-migrants who are looking for work and welfare.

Many of the urban consequences of the absorption into the global economy are exemplified by Zimbabwe, a country which attained formal sovereignty in 1980 after 15 years of unilaterally declared independence (Drakakis-Smith, 1992). The modern urban system in Zimbabwe emerged under settler colonialism to facilitate the export of various commodities and the import of consumer goods. Cities were dominated by the white minority in the country, and other than those employed in domestic service and a very small number in industry and service activities, blacks were prohibited unless they had a job and accommodation. In the countryside, some blacks worked for white farmers but most were engaged in subsistence agriculture. The population was 17 per cent urban in 1970. The favouring of the white colonialists, however, meant that social and health care services were city-based and significant differences in standards of provision existed between urban and rural areas.

This basic pattern was transformed during the 1970s as a consequence of increased foreign investment and the opening up of external markets for the products of Zimbabwe's farms and factories. Urbanisation occurred through net in-migration to jobs in cities as the manufacturing sector increased its contribution to the gross national product from 10 per cent in 1965 to 24 per cent in 1980 (Stoneman, 1979). At the same time, the mechanisation of many of the larger commercial farms, and their increase in size, generated a surplus of black labour in rural areas. Movement into the cities increased significantly after 1980 when the legislation controlling ownership and residence in cities was relaxed and removed. Many traditionally white areas of Zimbabwe's cities rapidly became black (Cumming, 1990). Urban growth was compounded when families were reunited and birth rates rose. Some 31 per cent of the population are thought to live in urban places in 1995 and the population of Greater Harare is in excess of 1.5 million. The recent rapid urbanisation in Zimbabwe, in common with many African and Asian countries, is a consequence of structural and associated spatial changes which are associated with the transformation of a rural subsistence economy into an urban-based and politically independent commercial economy which is incorporated within the global economic system.

Work on migration patterns in China by Goldstein (1993), however, shows that urbanisation is taking place in parts of the periphery for reasons which are largely unconnected with the emergence of the world-economy. With the exception of a small number of economic sectors, China largely functions outside the world system and yet it has undergone major and rapid urban development in recent years. Evidence from national population surveys analysed by Goldstein suggests that urban growth and urbanisation are principally products of migration, with three-quarters of all population movements being from rural to urban areas. This is despite the operation of a strict residential registration system which attempts to control internal migration. Because of the difficulties of re-registering, the official statistics exclude large numbers of migrants who further swell the populations of China's principal cities. The number of unofficial migrants in the 23 largest cities in 1990 was estimated to be around 10 million. They are thought to number in excess of 1 million in Beijing alone.

Urban growth and urbanisation in China are principally a consequence of domestic economic and social circumstances. The growth of population in the countryside, together with the economic reforms which were introduced after 1979, created a vast surplus of labour which could not be absorbed by the already over-extended rural economy. Cities are powerful magnets for displaced peasants whose annual incomes depended on the vagaries of the weather and until the 1980s on income distribution decisions by the collective leadership of the communes. Moreover, economic liberalisation allows and indeed encourages farmers to market their products in cities. Social advantages associated with a movement to cities include the opportunity to escape from restrictions on marriage, size of family, and the burdens of having to look after elderly parents. Urban development in China is largely occurring from within as a consequence of the adjustment from a rigid and traditionally repressive rural subsistence to a more open and liberal service-based urban economy.

Detailed evidence on the precise nature of the links between global production and urban development in the periphery is presently fragmentary both because the relationship is new and because the data on recent urban growth in the world's least developed and poorest countries are lacking. The true scale and significance of urbanisation in the periphery have only recently become apparent. The research which has been undertaken points to the existence of a general relationship between investment in manufacturing, as a consequence of global restructuring, and urban growth but with wide variations from country to country. Taiwan, Singapore and Korea, where there is a clear connection, and China, where urban development is largely a consequence of rural changes associated with economic liberalisation, perhaps represent the extremes. In seeking explanations for urban development it is important also to distinguish within the periphery between experiences in South America, where levels of urban development are historically high, and Africa and Asia, where they are low. Both types of area have been affected by the same shifts of production activity associated with the emergence of the new economic order, but with different consequences. The effects on cities and urban systems in the former have largely been to consolidate existing patterns by compounding growth in existing centres. In Africa and Asia they have created and accelerated urban development where little existed before.

SOCIO-ECONOMIC CONSEQUENCES OF GLOBAL URBANISATION

It is important to emphasise that urbanisation in Africa and Asia is presently taking place in countries with the lowest levels of economic development. This is the opposite of the historical situation, when accelerated urbanisation began in the advanced economies of Western Europe and North America, and it has far-reaching implications for the ability of national governments to cope with its profound social consequences (Berry, 1973). Contemporary urbanisation involves countries in which the people have the lowest levels of life expectancy at birth, the poorest nutritional levels, the lowest energy consumption levels and the lowest levels of education. It involves far greater numbers and is taking place over a much wider geographical area than was the case, historically, in the core. The causes are also different and owe little to indigenous development, being largely a local consequence of changing economic and political relationships at the global scale. Industrialisation lags well behind urbanisation, so that many people in the city, and especially recent in-migrants, find at best marginal employment in small-scale production and service activities.

The emergence of an unorganised, unregulated and unregistered informal sector in urban economies in many developing countries is a widespread corollary of contemporary urbanisation. Its existence was recognised and documented in the early 1970s when it was observed that massive additions to the urban labour force, through high net in-migration and natural increase, were not reflected in the employment statistics. Most people in the informal sector work for themselves or for small-scale family-owned enterprises. The self-employed are engaged in activities ranging from hawking, street vending, letter writing, shoe cleaning, knife sharpening and junk collecting to prostitution, drug peddling and snake charming. Others work as mechanics, carpenters, artisans, barbers and personal servants. Some run successful small-scale enterprises with several employees (Torado, 1994: p. 253). Survey work by Sethuraman (1981) in thirty-two cities in the developing world suggests that as many as half of the urban workforce are typically employed in informal sector activities. The size and relative importance of the sector are expected to increase significantly as many more people are added to cities in the developing world than can find work in formal employment.

Although its existence may reflect shortcomings in the urban economy, it is important to emphasise the many positive aspects of the informal sector. The first and most obvious is that it provides jobs and income for large numbers of people who otherwise would have no means of economic support in countries which lack basic welfare services. Many are escapees from rural poverty and although living conditions in the city are little better than those in the countryside, the sector gives them first access to an urban economy in which there are opportunities and potentials. Work in the informal sector is undertaken by women and children to the benefit of the family. A disproportionate number of workers are in fact women, who find it especially difficult to get jobs in a formal sector which is generally dominated by men.

The informal sector is valued for its vitality and dynamism. Despite very low levels of turnover some surpluses are generated, which are especially valuable in developing countries where there are shortages of capital. It is a sector in which jobs are created at low, and in some cases no, cost, where few skills may be required and where elementary instruction, training and experience can be gained. The informal sector is more likely to adopt and develop appropriate technologies and make effective use of resources, and it plays an important role in the recycling of waste materials, many of which find their way into the industrial sector or provide basic commodities for the poor. Against these benefits must be set the consequences of low standards of health, safety and hygiene, the low standards of living which are supported, and the ease with which vulnerable groups can be exploited. Despite these negative aspects the informal sector is the lifeline for perhaps one billion people in cities in the developing world and the opportunities which it affords are major reasons for in-migration. Its existence is both a consequence and a cause of the current rapid urbanisation across large parts of the periphery.

Much of the rapidly growing urban population in developing countries is accommodated in shanty towns. Such self-erected and often illegal developments form highly distinctive features of the built environments of most major cities in the developing world. Shanty towns represent individuals' responses to acute need. Few cities can provide sufficient housing, and in-migrants would be unable to pay for it if it was available, so the population is forced to erect and live in makeshift dwellings. Those who are unable to find space or to accumulate the necessary resources and materials to construct a permanent shelter must live on the streets, and this is the fate of many in developing world cities. The *favelas* of Rio de Janerio, the *pueblos jovenes* of Lima, the *bustees* of Calcutta and the *bidonvilles* of Dakar are merely the best-known and most well documented of the squatter communities that are estimated to accommodate over one-third of the urban population in developing countries (Torado, 1994: p. 250).

Shanty towns have many features in common. They consist of closely packed constructions of mud, thatch, timber or corrugated iron which are erected on any available space without planning permission or building regulations approval. Many are on government-owned land, from which eviction is politically impossible. Residential densities are extremely high and drainage, sanitation and water supply are lacking or are deficient. Many developments are alongside factories, railway yards and public buildings which present opportunities for casual employment and informal acquisition. Most occupants are poor and work in the informal sector.

Despite their present importance, shanty towns are recent additions to the cities of the developing world. The earliest is reckoned to be the *favela* which was built in Rio in 1910, but the vast majority are products of the major and rapid urban growth and urbanisation of the population which has occurred since mid-century (Dwyer, 1979). The example of San Martin, Buenos Aires, illustrates the way in which shanty towns are created (Hardoy and Satterthwaite, 1989: p. 12). The research which they report describes in detail the activities of September 1981 when a small, well organised group of squatters invaded and occupied 211 hectares of abandoned private land. As word spread, some 3,000 people entered and established themselves in the settlement in a period of five days. Government efforts to bulldoze the area were resisted by women and children who stood in front of the machines, and attempts by the police to stop construction were thwarted when cordons were broken at night. At its peak the rate of building was staggering, with five or six houses being erected each day. By October 1982 the settlement was home to some 20,000 people. Although it was erected illegally the development was planned, with space being left for access roads and community facilities. A democratic system of organisation and administration soon developed. Conditions in San Martin remained poor because the local authority refused to pave streets, install sewers or provide health care, so the residents were forced to make their own arrangements. By 1984 a health centre and a school had been built.

The speed of development and ubiquitous nature of shanty towns mean that they are the focus of wide interest from analysts, planners and politicians. Attitudes vary from outright opposition to benign tolerance. The cramped conditions, poor construction and the fact that they constitute a visual embarrassment to municipal authorities and high-income groups mean that they can easily be viewed in negative terms. Shanty towns house populations at very high densities with increased risks to health, and commonly lack basic utilities and amenities. Houses are built to low standards out of materials which come to hand and provide inadequate protection from the elements. They are highly susceptible to fire and harbour disease-carrying rodents and insects (World Health Organisation, 1991). Many shanty towns are located in high-risk sites and are especially vulnerable to floods, landslides and the hazards associated with transport and with industry. Their very existence draws attention to the failings of legal processes, planning powers and government policies.

Many observers, however, view shanty towns in more positive terms since, as the San Martin example shows, they can emerge as vibrant communities in their own right with strong forms of social organisation. Shanty towns can be an important first destination for new arrivals to the city and help·them to become familiar with and assimilated within urban life. They provide valuable accommodation for those who would otherwise be without shelter, and their construction involves considerable reuse of materials. They provide a pool of cheap and accessible labour for urban industries. They are well established across the cities of the developing world and will not go away.

For these reasons some governments have adopted a pragmatic approach and have moved away from policies of opposing and demolishing shanty towns. Instead they seek to provide basic services and to help residents to upgrade their housing. The critical step in this process is the granting of legal rights to land, this enabling occupants to participate fully and openly in the economic and political life of the city (HABITAT, 1987). Shanty towns house large numbers of people and their votes are an important factor in municipal and national politics.

Shanty towns may look unattractive and be condemned by many governments, but they fulfil a vital social need. They symbolise the vigour and spirit of self-help, born out of necessity, that lie behind the rapidly expanding cities of the developing world. Shanty towns evolve through a sequence of illegal occupation and building, initial antagonism and opposition from the municipal authority, official acceptance and recognition, and, finally, formal incorporation into the wider city. This mechanism is the principal way in which urban growth is occurring in developing countries. Whether it can accommodate the expected increases in urban populations in the future and so produce sustainable cities is a central theme of Chapter 8.

CONCLUSION

This chapter has identified and has attempted to account for the emergence of urban growth and urbanisation, since mid-century, as a global phenomenon. It is sometimes difficult for analysts who view the world from universities in the historic cities of Britain and Europe to appreciate that urban development elsewhere is so recent and indeed in some countries has yet to begin. In some parts of the world cities themselves are a novelty. Although the origins of global urbanisation lie in mercantilism and industrial capitalism, the contemporary urban world is essentially a product of processes of population concentration which began to work powerfully in the core economies as late as the early part of the twentieth century, and have only become truly world-wide in extent and effect in the last twenty years. Urbanisation, when viewed at the global scale, is a contemporary phenomenon which, in being driven forward by transnational corporate capitalism, owes comparatively little to history. The same generalisations apply to the location of the urban population. Urban settlements have existed since the days of ancient Egypt and Mesopotamia, but many major cities, and nearly all mega-cities, are effectively less than half a century old.

Urban growth and urbanisation have transformed the global pattern of settlement over the past half century. A world that is today predominantly urban has replaced one in which most people, in 1950, lived in rural areas. The impact however has been far greater than one of mere location of population as towns and cities represent and have introduced new and different forms of social organisation, interaction and behaviour. Life in the city has a different purpose, rhythm and variety to that in country areas and affords a wide range of choices and opportunities. As well as shifting into urban places, the population is being urbanised in a social sense. The characteristics of urbanism, and the nature of the lifestyles which exist in cities, are explored in the next chapter.

RECOMMENDED READING

Auty, R. (1995) *Patterns of Development: Resources, Policy and Economic Growth*, London: Arnold.

A detailed analysis of recent economic performance in the developing world in which urban issues, including primacy and hyper-urbanisation, are evaluated in the context of economic and social change.

Chase-Dunn, C. (1989) *Global Formation: Structures of the World-Economy*, London: Blackwell.

An advanced discussion and evaluation of the emergence of global economic and social structures and relationships and their implications for cities and societies.

Drakakis-Smith, D. (1992) *Urban and Regional Change in Southern Africa*, London: Routledge.

A valuable collection of papers on contemporary urban change in southern Africa with useful case studies on selected countries.

Hardoy, J. E. and Satterthwaite, D. (1989) *Squatter Citizen*, London: Earthscan.

A detailed examination of the lives and prospects of residents of the squatter settlements, and informal sector workers, in Third World cities.

Kasarda, J. D. and Parnell, A. M. (1993) *Third World Cities*, London: Sage.

A useful overview of levels of contemporary urban development across the Third World, including detailed case studies of urban issues in selected countries.

5

LIFESTYLES IN THE CITY: TRADITIONAL PLACEBOUND INTERPRETATIONS

The progressive shift of the world's population from rural into urban places is accompanied by profound and far-reaching changes in the ways in which many people live their daily lives. Cities are different in physical, social and economic terms. They offer their residents a far wider range of options and opportunities and enable them to engage in many more interests and activities than are possible in rural areas. They are places with large numbers of people, factories, offices, shops and recreational facilities which facilitate, support and promote a variety of lifestyles which are distinctively urban in character and which differ fundamentally from those which occur elsewhere. Traditionally, such patterns of association and behaviour were thought to be a simple function of place, being restricted to and experienced by those who actually lived in the city. Today, long distance travel, telecommunications and the mass media extend urban influences and options well beyond settlement boundaries. Most people, wherever they live, are being exposed to urban attitudes and values as urbanism penetrates deep into rural regions. The ability to participate in an urban way of life is largely independent of location and is open to all. The world is increasingly becoming a global urban society of which we are all residents.

'Urban' is a descriptive label which is used to describe both a particular type of place and a set of distinctive patterns of association, values and behaviour. It is this latter, sociological meaning, which is addressed and examined in this chapter. The concept of urbanism is that of a set of lifestyles which arise in cities and follows from their impact on society. It is expressed and reflected through patterns of social and economic relationship and behaviour, and through taste, fads, fashions, aspirations and achievements. Urbanism is the least advanced of the principal processes of global urban change and it is by far the most difficult to conceptualise, document, interpret and explain. Urban growth and urbanisation have created towns and cities which house a little over half of the world's population, but the progress and extension of urbanism, however defined, is much more limited. Many people who live in remote locations have lifestyles which are similar to those in cities. They participate fully in an urban culture and their attitudes and values are the same as those of residents of the world's principal cities. Conversely, there are a large number of people, especially in the million and megacities of the developing world, who retain patterns of association and behaviour that are more akin to those in rural areas. Their attitudes and values are permeated and constrained by traditional influences of religion, the family and geographical parochialism. The lifestyles of many first- and second-generation in-migrants have not yet been affected by incorporation within urban society.

When viewed in this way, urbanism, like urbanisation, is seen to be a finite process. It begins with a wholly rural society in which urban influences are absent and ends when everyone everywhere lives an urban way of life. Mapping its advance is necessarily imprecise since it involves evaluating changes in attitudes and behaviour which are not capable of precise specification and measurement. Cross-national data on urban growth and urbanisation are available though of limited quality, but there are no comparable sources on lifestyles and behaviour. Most insights into urbanism are the product of in-depth sociological studies, of which there is a deficiency in developing countries. Such work as does exist needs to be evaluated critically since it is difficult if not impossible for researchers to shed their own cultural values and so study alien societies, and indeed their own, objectively and dispassionately. What it suggests, however, is that society in Western Europe, much of North America and parts of Australasia, the Middle East and South America is deeply permeated by urban values and dominated by urban institutions, but elsewhere the incidence of urbanism is limited. Vast tracts of Africa and Asia, and extensive areas in other parts of the developing world are largely unaffected by urban influences. Rural ways of life predominate over large parts of the contemporary urban world.

Urbanism spreads through indigenous change and spatial diffusion. At the simplest level it can be argued that urban lifestyles originate in cities in response to the opportunities and constraints afforded by place. From there they extend outwards to surrounding rural areas. Place influences and conditions behaviour because people in cities live in large numbers and at high densities in artificial environments and so they develop different patterns of association and living. Urbanism spreads through direct contact to adjacent rural areas and hierarchically, through the settlement system to smaller towns. The values and forms of behaviour which originate and diffuse through indigenous change are likely to be culturally conservative and so will fall within locally acceptable bounds of 'progress' or 'modernisation'.

Urbanism can, however, be extended and exported to distant destinations well beyond the city via the media of print, film, tape, disk, radio and television. The introduction of satellite relays, long-distance telecommunication, audio and video cassettes, and the Internet, has enabled the reach of these media to become global. They can be accessed over a large part of the earth's surface by anyone who has access to the basic technology. Such material is typically loaded with urban images, symbols and values and popularises and promotes urban patterns of social and economic relationship. Such media portray and project lifestyles and modes of behaviour in cities which may differ radically from and so clash fundamentally with those which are the accepted norm in traditional societies. The world's rural areas are being exposed to, and are being forced to come to terms with, urban influences both from within their own countries and from societies which may be both culturally as well as geographically distant.

Research into the characteristics and geographical spread of urbanism has a long lineage. It traces its origins to sociological studies of the industrial cities of the early twentieth century in which distinctive urban lifestyles were first recognised and documented in detail. This work identified the existence of close links between the social and economic characteristics of cities and the patterns of activity and behaviour within them, which led researchers to conclude that it was principally location that determined an urban way of life (Table 9). On this basis, urbanism was believed to increase with urbanisation as people who moved into cities adopted urban lifestyles in place of rural patterns of behaviour. With the increasing sophistication of society, however, it was observed that individuals who live in cities enjoy access to a wider range of lifestyles. These are selected according to the circumstances of the individual, with class, ethnicity and age being especially important influences. Lifestyles were matters of choice rather than compulsion, the degree of choice being widest in big cities where the range of opportunities is greatest.

Table 9 Alternative explanations of urban lifestyles

	Leading proponents	Key variables	Effects of urban life on social groups	Socio-psychological consequences
Determinist	Wirth (1938)	Size, density and heterogeneity	Breakdown of primary groups	Alienation, deviance, anomie
Compositional	Gans (1962a, 1962b); Pahl (1965)	Class ethnicity, stage in life cycle	No direct effect	No direct effect. (Indirect effects as a consequence of class, ethnic and lifestyle position)
Subcultural	Fischer (1976)	Size, 'critical mass', interaction	Creation of primary groups	Subcultural integration

A principal benefit of cities is indeed that because of their size, critical mass and the opportunities for interaction, they create and sustain a rich variety of subcultures and so provide their residents with a wide range of lifestyle options.

Urban values and attitudes became independent of the city with the development of the mass media. As such they are elements in the progression to modernity that is affecting society generally (Giddens, 1991). Urbanism in an increasingly aspatial world affects not merely the immediate surroundings of the city, but distant and remote regions both within the country and across the world. The rapid spread of urbanism over the last two decades via film, television and video was viewed with alarm by some observers, who raised the spectre of urban images, attitudes and values flooding out of decadent Western cities to the detriment of ancient and rich rural cultures in distant developing countries, a vision that was readily conceptualised within the framework of 'cultural imperialism'. Recent evaluation of the evidence, however, suggests that this model is too simple and that urban lifestyles are being extended across the globe within the framework of cultural pluralism; rural areas in developing countries are being exposed to and are absorbing values and attitudes from many diverse urban sources both locally and in the developed world. This chapter explores the bases and characteristics of urban lifestyles by examining traditional placebound interpretations, while Chapter 6 explores the extension and spread of urban values in rural areas by means of the mass media. Together they seek, from an early point in what is expected to be a long-term process of fundamental social change, to document and explain the origin and diffusion of urban lifestyles from the industrial metropolis to the projected global urban society.

URBAN WAYS OF LIFE

Attempts to describe and explain the characteristics of urban society are most closely associated, historically, with the work in the 1920s and 1930s of the Chicago School of Human Ecology. Robert Park, the founder of the school, believed that the best method of studying the new urban ways of life was for social scientists to go out and explore their own cities. Unparalleled opportunities for field observation were provided by Chicago which, in the first two decades of the twentieth century, and like many cities in developing countries today, was expanding rapidly through immigration, and exhibiting all of the stresses and tensions associated with explosive growth. On this basis, a large volume of research developed which provided a wealth of empirical data. One of the last members of the school, Louis Wirth, was responsible for drawing all this work together in a seminal essay entitled 'Urbanism as a Way of Life' (1938).

Wirth set out to discover the forms of social action and organisation which typically emerged when people congregated in cities. Specifically, he identified the three dominant characteristics of the city as being its large size, its dense population concentration and its heterogeneous social mix. For Wirth, the size of the social group determined the nature of human relationships. Increase the number of inhabitants in a community beyond a certain level, and the possibility of each member knowing all the others personally is reduced. Moreover, urban dwellers do not become involved with one another as total personalities, but in specialised segments, interaction being for specific reasons. Under these circumstances, individuals tend to form only weak links with others, so that the close bonds of family and neighbourliness, present in folk cultures, give way to differentiation, specialisation and symbolism. Formal methods of social control replace informal methods, and indirect modes of communication replace personal contacts. The value of social relationships is measured in monetary terms, and will be manipulated as a means of achieving one's own ends. The effect is that the individual will come to count for little and will be subsumed as an anonymous member of a social group, responding to institutionalised codes of behaviour. Lacking the security provided by familiar norms and sanctions, the individual will feel a sense of personal detachment and disorganisation.

As density of population increases, so area specialisation results. The competition for space becomes so great that each area in the city tends to be put to the uses which yield the greatest economic return. Place of work becomes divorced from place of residence, and place and nature of work, income, habit, taste, preference and prejudice combine to produce a matrix of social worlds in the city. The close living together and working together of individuals who have no sentimental ties or emotional links fosters a spirit of competition, aggrandisement and mutual exploitation. For those unable to find a secure life in some specialised role or sub-area, the likelihood of dysfunctional behaviour increases, especially where densities are highest.

The heterogeneous nature of urban populations, Wirth believed, led to social instability and personal insecurity. Individuals acquired membership of widely divergent social groups, each of which functions with respect to a single segment of their personality. Geographical and social mobility mean that turnover in group membership is rapid, so this further limits the possibility of establishing intimate and lasting relationships. As a consequence of transitory habitat, there is little opportunity for individuals to obtain a conception of the city or to survey their place in the total scheme. Detachment from the organised bodies which integrate society leads to depersonalisation and social isolation.

Wirth argued that the large, dense and heterogeneous concentrations of population that occur in cities demand a response from the individual. The crowding of diverse types of people into small areas broke down existing social and cultural patterns. He concluded, in very negative terms, that sooner or later the pressures exerted by the dominant social, economic and political institutions of the city would destroy primary group relationships and all the forces of social control which are derived from them. The end product is 'anomie', a condition in which the normal rules and conventions which regulate social behaviour break down, 'alienation' in which the individual becomes detached from society, and social 'deviance'.

Although the principal aim was to account for the variety of lifestyles in cities, Wirth's prediction of a progression of social and behavioural differentiation between the urban and rural formed the basis of a powerful geographical model that was of potential relevance at the global scale. At one extreme was the metropolis where, because of large, dense and heterogeneous concentrations of population, urban ways of life were expected to be most clearly apparent; at the other, in traditional folk societies, rural patterns of interaction, association and behaviour, untouched by urban influences, would be found. Between these two ends of the spectrum, gradations of social and behavioural differentiation, reflecting the level of urbanism, were postulated. Within three years of the publication of Wirth's paper, Redfield's study of the *Folk Culture of Yucatan* (1941) claimed to provide empirical support for this contention. Redfield analysed in depth the social and economic characteristics of a tribal village (population 250), a peasant village (population 250), a town (population 1,200) and a city (population 100,000) which were held to be representative of the range of settlements in Yucatán. His most important observation was that there were regular and progressive variations among them, from the city of Merida at one extreme to the isolated settlement of Tusik at the other, and these differences, it was argued, were indicative of the existence of a continuous scale of rural–urban development:

> the peasant village as compared with the tribal village, the town as compared with the peasant village, or the city as compared with the town is (1) less isolated; (2) more heterogeneous; (3) characterised by a more complex division of labour; (4) has a more completely developed money economy; (5) has professional specialists who are more secular and less sacred; (6) has kinship and godparental institutions that are less well organised and less effective in social control; (7) is correspondingly more dependent upon impersonally acting institutions of social control; (8) is less religious, with respect to both beliefs and practices of Catholic origin as well as those of Indian origin; (9) exhibits less tendency to regard sickness as resulting from a breach of moral or merely customary rule; (10) allows a greater freedom of action and choice to the individual.
>
> (Redfield, 1941: pp. 338–9)

LIFESTYLE STUDIES

Following Redfield's work, a wide range of studies looked for evidence of progressive differentiation among settlements along the rural–urban continuum. Particular attention focused on the ways of life of the residents of the central area of the metropolis. It is here that Wirth conceived the urban population as consisting of heterogeneous individuals, torn from past social systems, unable to develop new ones, and therefore prey to social anarchy, crime and deviance. Research, however, failed to uncover evidence to substantiate this wholly negative view of the city. Indeed all the indications are that most cities are successful and stable. The principal reason is their social and economic variety and character which is reflected in the richness of cultural forms in the city. In place of a single response determined by size, density and heterogeneity of the social group, central areas in particular offer a wide range of mutually supporting social and behavioural milieux.

Lifestyles in central cities were researched in detail and were codified and explained in a number of studies undertaken by Gans and others in the 1960s. For Gans (1962a) there are five basic types of inner city resident: the urban villagers, the cosmopolites, the unmarrieds or childless, the trapped and downward mobile and the deprived. The most well documented are the urban villagers, so-called because they are members of small, intimate and often ethnic communities based upon interwoven kinship networks and a high level of primary contact with familiar faces. The characteristics of urban villagers and their ways of life were described by Gans in his study of the residents of Boston's West End (1962b). The area was populated by immigrants from a variety of national and ethnic backgrounds including Italians, Poles, Irish, Greeks, Ukrainians, Albanians and Jews. Low incomes, limited occupational skills and qualifications and poor housing were common features. Behind the somewhat offensive facade of the area, which was strongly influenced by the dilapidated state of the buildings, the vacant lots and the garbage on the streets, Gans found a friendly and close-knit community which reminded him of that which exists in small towns and rural areas. Far from diminishing in importance, the family remained a major component in social organisation, and religion retained its hold upon the people. The sharing of values was also encouraged by residential stability and the diverse network of personal acquaintances. Everyone might not know everyone else, but as they did know something about everyone, the net effect was the same, especially within each ethnic group. Between groups, common residence, sharing of facilities and the constant struggle against absentee landlords created enough solidarity to maintain a friendly spirit. Although for many families problems of unemployment, finance, illness, education and bereavement were never far away, neighbours and friends were always on hand to provide assistance and support. The relevance of Gans' generalisations may be criticised on the grounds that the West End was populated by predominantly first-generation immigrants who had yet to be exposed to the full impact of urban living, but this was not the case in Bethnal Green, London (Young and Willimott, 1957) where similar urban village communities were identified. Together, these studies point to a way of life in central areas which differs sharply from Wirth's urbanism. Far from being reduced, life in the urban village is organised around kinship and primary groups which protect the individual from deviance, alienation and the descent into anomic forms of behaviour.

Gans' second group, the cosmopolites, includes students, artists, writers, musicians and entertainers, as well as intellectuals and professionals who live in the city in order to be near the special cultural facilities of the centre. Many are unmarried or childless, but others rear children in the city, especially if they have the income to support the services of au pairs or childminders. Still others, though having an out-of town residence, maintain a cosmopolitan lifestyle from a *pied à terre* in the city. The unmarrieds or childless are divided into two sub-types, depending on the permanence or transience of their status. The temporarily unmarried or childless live in the inner city for only a limited time and upon marriage or starting a family, they leave for the outer city or suburbs; the permanently unmarried live in the inner city for the remainder of their lives, their housing depending upon income. The former typically includes students or young people who share a downtown apartment or flat away from parents but close to jobs or educational opportunities.

The fourth group are the trapped and downward mobiles, who are people who stay behind when a neighbourhood is invaded by non-residential land uses or by lower-status immigrants, because they cannot afford to move, or are otherwise bound to their present location by tenancy or leasing agreements. Those who started life in a high-class position but who have been forced down in the socio-economic hierarchy and in the quality of their accommodation because of personal circumstances are typical members of this group. It may also include old people living out their existence on small pensions. The final group is the deprived population: the very poor, the emotionally disturbed or otherwise handicapped, broken families and non-whites. They are restricted by the housing market to dilapidated properties and blighted neighbourhoods although among them are some for whom the slum is a refuge or temporary stop-over to save money for a home in the outer city.

There is no suggestion that these types are mutually exclusive or that they are present in all central areas. Gans' work highlights the variety of lifestyles that may be followed by residents of the central city. Members of these different groups have such diverse character-istics and ways of life that it is hard to see how density and heterogeneity could exert a common controlling influence. The cos-mopolites and the unmarried and childless live in the inner city because they wish to, the urban villagers are there partly because of necessity, partly because of tradition. The final two types are in the inner city because they have no option. For Gans, lifestyles in the inner city involve some element of choice which in turn is heavily conditioned by background and class; they are independent of location.

A second major focus is upon lifestyles in the suburbs. As an area of the city, the suburbs date from the inter-war period in the United Kingdom and from the immediate post-war era in North America, when large-scale residential development at what was then the edge of the city took place. The most important feature of residential construction was that comparatively few designs were employed, so that the suburbs are characterised by streets or avenues of proper-ties of broadly similar size and style. Muller (1981) identifies four basic community forms within the contemporary American suburb: the affluent/exclusive areas; the middle-class family area; low-income ethnic-centred working class communities; and cosmopolitan centres. These areas differ markedly in terms of incomes, age structures, population stability and education, and are typified by lifestyles which vary according to patterns of association, parochialism and social mobility.

The high-income suburbs are characteristically located in areas which possess both physical isolation and the choicest environmental amenities around water, trees and higher ground. Since houses are built on large plots and are well fenced, neighbouring is difficult, and people keep in touch by formal participation in local social networks. These are tightly structured around organisations such as churches, country and golf clubs, and newcomers to the commu-nity are carefully screened for their social credentials before being accepted. Exclusiveness is reinforced by private schools and by the emphasis placed on class traditions. A recent development in high-income areas is the growth of luxury apartment and condominium complexes which attract increasing numbers of affluent singles and senior citizens.

Middle-class family suburbs are located as close as possible to the high-status residential enclaves of the most affluent. They are populated by middle-income groups which are typically arranged into nuclear family units. The rearing of children is the central concern and much local social contact occurs through family-oriented organisations such as school associations and children's societies and sports clubs. Despite the closer spacing of homes and these integrating activities, middle-class suburbanites are not communally cohesive to any great degree. Emphasis upon family privacy and the freedom aggressively to pursue upward social mobility does not encourage the development of strong social ties. Neighbouring, which is mostly child-related, is limited and selective, and even socialising with relatives is infrequent. Most social interaction revolves around a non-local network of self-selected friends, widely distributed across suburban space. The insular single family, detached family house and dependence upon the automobile for all trip-making accommodate these preferences and link lifestyle to the spatial arrangement of the urban environment.

Outside the central city, the working-class and poor areas are to be found in the innermost pre-automobile suburbs and adjacent to industrial areas and railroads. For Muller (*ibid.*: p. 71), working-class suburban lifestyles differ from middle-class suburban lifestyles in a number of ways. Whereas middle-class suburbs stress the nuclear family, socialising with friends, and a minimum of local contact, working-class neighbourhoods accentuate the extended family, frequent home entertaining of relatives and considerable informal local social interaction outside the home. The latter is reflected by the importance of local meeting places such as churches, taverns and street corners. These informal contacts provide an important element of social cohesion in working-class suburbs. Moreover, local area attachment is reinforced by a person-oriented rather than by a material or achievement-oriented outlook. Working-class suburbanites have limited opportunities for social advancement and so view their present home and community as a place of permanent settlement.

Muller's fourth category of suburban lifestyle is the suburban cosmopolitan centre. It is distinguished by communities of professionals, intellectuals, students, artists and writers who participate in far-flung intra- and inter-metropolitan social networks and communities of interest. As theatres, music and arts facilities, fine restaurants and other cultural activities have deconcentrated, so cosmopolitan centres have spread throughout the city, and lifestyles that were once the preserve of inner areas have become predominantly suburban. This expansion has been assisted by the opening of branch campuses of universities and colleges which provide a cultural and intellectual focus in the suburbs.

The very different types of community and associated lifestyles that can be recognised in the contemporary American suburbs suggest that location is unimportant as a controlling factor. What differentiates suburban communities is occupation, social class and ethnicity, and these are largely independent of size, density and heterogeneity considerations. Especially important are considerations of income, which determine both residential location and patterns of social interaction. The suburbs, like the central city, offer a range of distinctive niches within a mosaic culture which is increasingly dominating urban America and many other parts of the developed world. Lifestyles are not determined by position within a rural–urban continuum, rather they reflect the socio-economic characteristics of the population.

A broadly similar set of conclusions arise out of studies of lifestyles in the urbanised fringes of 'rural' areas around large cities. In view of the intermediate position along the continuum, the population of commuter zones may be expected to exhibit features of both urban and rural ways of life. Pahl's (1965) study of the northern commuter belt of London identified the major groupings in the representative commuter village and explored their lifestyles. They include large property owners, the salariat, retired workers with some capital, urban workers with limited capital or income, rural working-class commuters, and traditional ruralites.

The first group is the large property owners. They are tied closely to the village by tradition and land holdings, but may have considerable financial and business connections elsewhere. Amounting perhaps to only one or two families in each village, they may be absent for long periods and so play no real part in daily village life. The professional and managerial salariat choose to live in a commuter village not merely for the physical surroundings, but also on account of the distinctive pattern of social relationships which they associate with rural communities. They are attracted by the perceived 'friendliness' of village residents, although the degree of mixing, especially with manual workers and their families, is limited. The salariat and the manual workers commonly send their children to different schools, perform different roles in local village associations, and hold

111

contrasting opinions over local issues such as conservation and rural traditions. Retired urban workers with some capital are a third group who choose to live in a village because it is perceived to offer an attractive locality in which to spend old age. Because of age, family ties and financial independence, they participate in a distinctive pattern of social interaction, much of it non-village based.

Pahl's fourth category is that of urban workers with limited capital and income and includes those who are compelled on financial grounds to seek cheap housing in the metropolitan village. Although reluctant commuters, they depend heavily upon local services and so integrate comparatively readily into the village community. In contrast to this group, the rural working-class commuters are traditional village residents who find that their work is outside the village. The nature of employment centralisation in many rural areas suggests that this is an expanding group. The final category, that of traditional ruralites, comprises a small minority of local tradesmen and agricultural and related workers whose residence and employment are both local. There may be close kinship and other ties with the rural working-class commuter and in practice it is difficult to distinguish them in sociological terms.

Despite a common location, Pahl's analysis shows that lifestyles in the metropolitan village are highly differentiated. As in the inner city, size, density and heterogeneity of population do not impose a common social or behavioural response. 'There are some people who are in the city but are not of it (the urban villagers), whereas others are of the city but are not in it (the mobile middle class of the metropolitan village)' (Pahl, 1968: p. 273). Rather than being determined by location, lifestyles appear instead to be a function of constrained individual choice.

Proponents of a compositionalist theory of urbanism argue that choice is a function of an individual's background and position in society. These characteristics influence both social and locational aspirations, and also the extent to which those personal goals will be achieved. In contrast to the deterministic approach as represented by Wirth, compositionalists do not believe that city living weakens small primary groups, rather they maintain that these groups carry on undiminished. In place of size, density and heterogeneity, an individual's behaviour is held to be a product of social class, ethnicity and stage in life cycle. These characteristics determine patterns of interaction and association and so give rise to a mosaic of social worlds based upon kinship, neighbourhood, occupation, education or similar personal attributes. Rather than being destroyed, these private milieux endure and flourish in even the most urban of environments.

Compositionalists do not suggest that living in cities has no socio-psychological consequences, but they do maintain that any direct effects are insignificant. If community size has any effect, it results from the way in which it influences the position of individuals in the economic structure, the ethnic hierarchy and the life cycle. For example, large communities may provide better-paying jobs, and the people who obtain them will benefit significantly, but they will be affected by the new economic circumstance and not by the urban experience itself. Similarly, the city may attract a disproportionate number of males, so that many of them cannot find wives. This will certainly affect their behaviour but not because the city has cut their social ties. Fischer (1976: p. 35) explains the difference between the determinist and the compositionalist approaches in this way:

> both emphasise the importance of social worlds in forming the experiences and behaviour of individuals but they disagree sharply on the relationship of urbanism to the viability of these personal milieux. Determinist theory maintains that urbanism has a direct consequence on the coherence of such groups with serious consequences for individuals. Compositional theory maintains that these social worlds are largely impervious to ecological factors and that urbanism has no serious direct effects on groups or individuals.

The contrasting viewpoints represented by the determinist and the compositional approaches have been synthesised by Fischer (1976) in the form of a subculture theory of urban lifestyles. Subculture theory contends that living in cities independently affects social life not by destroying social groups, as Wirth's deterministic approach suggests, or by leaving them untouched, as the compositionalists believe, but instead by helping to create and strengthen them. The most significant social consequence of community size is promotion of diverse subcultures (culturally distinct groups, such as college students or recent immigrants). Like compositional theory, subcultural theory maintains that intimate social circles persist in the urban environment. But like determinism, it maintains that ecological factors produce significant effects on the social ordering of communities precisely by supporting the emergence and vitality of distinctive subcultures.

113

Fischer's subculture theory holds that there are two ways in which urban mosaics are produced. The first is that large communities attract immigrants from a wider area than do small towns and so they receive a great variety of cultural backgrounds which contribute to the formation of a diverse set of social worlds. The second is that as predicted by determinists, large size produces a differentiation of occupational and social functions. Where population size is large enough, what would otherwise be only a small group of individuals becomes a vital, active subculture. Sufficient numbers allow them to support institutions such as clubs, newspapers and social functions which serve the group and allow them to have a visible and coherent identity. For example, if only one person in a thousand is interested in amateur dramatics, there would only be five in a town of 5,000 – enough to do little else but to engage in conversation about acting. But in a city of one million, the thousand interested individuals would be sufficient to support plays, to lay on special visits to the theatre and to maintain a special social milieu. Interaction between members of a social world and those outside it may serve to break down barriers, but a more common reaction is for a defensive retreat into one's social world so that the cohesion of the group is intensified and reinforced. Subcultures do not therefore exist because social worlds in the city break down under the impact of ecological forces, but rather because within large, dense and heterogeneous populations their formation is facilitated and encouraged.

CONCLUSION

Lifestyles in the city may take a number of forms. The values, expectations and aspirations of the farm labourer have little in common with those of the bohemian, the affluent suburbanite or the long-distance urban commuter, and yet all may reside comparatively close to one another, in and around the city. Explanations for the spread of urbanism begin with traditional rural lifestyles and seek to identify and account for their progressive replacement by urban ways of living. At first glance, it seems reasonable to associate such changes with increasing community size, and in particular with the effects of concentration in large, dense and heterogeneous groupings, but the range of responses to urban living seems too wide to be explained by simple locational determinism. What cities offer is a mosaic of social worlds characterised by many social and economic differences. The extent to which individuals are free to choose among these lifestyles is clearly dependent upon their own personal circumstances. It is widest in the case of high-income, professionally qualified mobile groups, but is non-existent for members of the poverty-stricken underclass.

The spread of urbanism amounts to social change on a vast scale. It means deep and irrevocable transformations that affect every aspect of social life and all sections of society. There is little doubt that such changes were initiated by the explosive growth of large cities that began in the late eighteenth century. Equally it seems clear that in advanced Western societies their effects and ramifications are extensive. Elsewhere the pattern is more varied and the fusion of urban and rural has yet to begin in the many countries of the developing world in which the population remains predominantly village-based. Not everyone who presently lives in cities, however, is urbanised, as there are many who exist outside or on the fringes of urban society. This is especially so in the rapidly growing million and mega-cities of the developing world whose social systems are unable readily to incorporate and absorb newly arrived migrants. Such people retain ways of life which are closely akin to those in rural areas. Conversely, there are many who live well beyond the city who nevertheless have lifestyles and patterns of behaviour which are essentially urban in character. The ways in which such rural residents are affected and conditioned by cities are considered in the next chapter.

RECOMMENDED READING

Berry, B. J. L. (1973) *The Human Consequences of Urbanisation*, New York: St Martin's Press.

An incisive overview and critique of the social correlates and consequences of urbanisation in the nineteenth and twentieth centuries in both the developed and the Third World. The section on Wirth's concept of urbanism and its relevance to the industrial metropolis is especially useful.

Chaney, D. (1994) *The Cultural Turn: Scene-setting Essays on Contemporary Social History*, London: Routledge.

A largely theoretical sociological examination of the links between place, culture and lifestyle.

Fischer, C. S. (1976) *The Urban Experience*, New York: Harcourt Brace Jovanovich.

A comprehensive review of alternative interpretations of the origins of urban lifestyles, with particular emphasis on the merits of the subcultural approach.

Muller, P. O. (1981) *Contemporary Suburban America*, Englewood Cliffs NJ: Prentice Hall.

An illustrative examination and analysis of the range and variety of suburban lifestyles in US cities and the reasons for them.

Walmsley, D. J. and Lewis, G. J. (1993) *People and Environment*, London: Longman.

A comprehensive introduction to the study of the relationship between location and behaviour, with useful sections on urban social patterns and structures.

6

GLOBAL URBAN SOCIETY

The discussion in the previous chapter focused on the characteristics and explanations of the patterns of association, interaction and behaviour that are commonly found in cities and how they differ from those in rural areas. Emphasis was placed upon location and how, by destroying, having no effect upon or creating primary groups, it influences social relationships and behaviour which are expressed in distinctive urban lifestyles. An important feature of much of contemporary urbanism, however, is that it has become independent of the city, both locally and at the global scale. It is no longer determined by or associated with place. Although individuals may live in remote areas in predominantly rural territories, many participate actively in an urban way of life. Access to urban lifestyles, once the prerogative of those actually resident in and around the city, is open to all. Urbanism, divorced from the city, is a principal element in a general process of modernisation.

The interest in the future of communities in a world in which space was losing its importance arose out of the growing realisation during the 1960s that transport and telecommunications had far-reaching implications for the organisation of society at the global scale. It was at this time that the introduction of jet travel and a big increase in the size and quality of telephone services undermined the frictional effects of time and distance and effectively shrank the world to a more manageable size. The urban consequences were first addressed by the communications guru Marshall McLuhan in his book *Understanding Media: The Extensions of Man* (1964). For McLuhan, history is made up of three stages. The first is the pre-literate or tribal stage, in which people live close together and communicate orally. This was followed by the Gutenburg or individual stage, in which communication takes place by the printed word, and thinking is done in a linear-sequential pattern. The third is the neo-tribal or electric stage, in which computers, television and other electronic communication media move people back together. The significance of the latter is that it makes clustering of activities and people at particular places unnecessary:

> Before the huddle of the city there was the food gathering phase of man the hunter, even as men have now in the electric age returned psychically and socially to the nomad state. Now however it is called information gathering and data processing. But it is global and it ignores and replaces the form of the city which has therefore become obsolete. With instant electric technology the globe itself can never again be more than a village and the very nature of city as a form of major dimensions must inevitably dissolve like a fading shot in a movie.
>
> (McLuhan, 1964; p. 366)

McLuhan's arguments echoed and extended those of Meier (1962) as set out in his communications theory of urban growth. Meier analysed the nature of the bond between two individuals and suggested that the quantum involved was the transaction; the exchange or transfer of information between sender and receiver. Bond formation between individuals, he argued, was facilitated by geographical proximity and by the acquisition and retention of knowledge so that cities evolved primarily as a means of facilitating interpersonal communication. An important feature, indeed a major attraction of city life, was the time spent in public and professional as opposed to private and family life, so that shared symbols and experience generate civic bonds which help to maintain and reinforce the cohesion of the city. Urban growth therefore takes place as cities develop a capacity to maintain and conserve information. The city is seen as an open communication system, resulting from and held together by a complex pattern of information exchanges.

Meier related urban growth not to changes in economic product or to the size of the social group, but to developments in communications technology. Social exchange in early settlements was primarily achieved by face-to-face contact and, with people attempting to maximise their chances of social interaction by locating close to the central zone of conflux, cities were both densely populated and compact. They were dominated by a small elite of priests and politicians who, through public meetings and assemblies, controlled the distribution of community information. With the progressive introduction of handwritten records, printing, publishing and broadcasting, communications technology complemented, and then largely replaced, face-to-face contact as the prime means of information dissemination. Modern cities are defined not simply in physical terms but as social networks in space, created, maintained and manipulated by a wide range of communications media.

The most important characteristic of such networks is that they are defined by involvement and not by proximity. Although individuals live in a particular place and participate in community life in and around that place, it is interaction rather than location that is of primary importance. Individuals are increasingly able to maintain contacts with others on an interest basis and so can be members of interest communities which are not territorially defined. Modern communication technologies open up a choice of lifestyles, once the prerogative of those who live in the city, to all. Propinquity is no longer a prerequisite for community.

For Webber (1964), the extent and range of participation in interest communities is a function of an individual's specialisation rather than place of residence. The more highly skilled the people are, or the more uncommon the information they hold, the more spatially dispersed are the members of their interest group and the greater the distances over which they interact with others. There is a hierarchical continuum in which the most highly specialised people are participants in interest communities that span the entire world; others, who are less specialised, seldom communicate with people outside the nation but regularly interact with people in various parts of the country; others communicate almost exclusively with their neighbours (Figure 20). For example, research workers will be members of a wide range of interest communities and will spend more of their time participating in national and international scale communities than will the primary school teacher. Similarly, the upper limit of the interest community of school caretakers will in all probability be restricted to the city. All three will, however, be involved in more parochial interest groups in association with their roles as parents, or members of local clubs and societies.

For any given level of specialisation, Webber argued that there is a wide variety of interest communities whose members conduct their affairs within roughly the same spatial field or urban realm. Such urban realms are neither urban settlements nor territories, but heterogeneous groups of people communicating with each other through

119

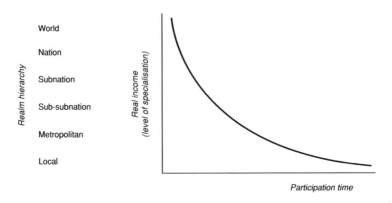

Figure 20   Webber's concept of the structure of interest
communities by realm

space. They are somewhat analogous to urban regions, but, contrasting with the vertical divisions of territory that are organised
around cities and accord with the place conception of regions, urban
space is divided horizontally into a hierarchy of non-place urban
realms (Figure 21). Irrespective of location, people at different
moments are participants in a number of different realms as they
shift from one role to another, but only the most specialised people
communicate across the entire nation and beyond. In this context
the city is no longer a unitary place. Rather it is part of a whole
array of shifting and interpenetrating realm spaces which exist at a
variety of spatial scales, from the local to the global.

Although one-third of a century has elapsed since its original conception, the non-place global village as envisaged by McLuhan, Meier,
Webber and others remains a distant prospect. Some people, most
notably those in the highest levels in business, government and acad-
emia, live communications-intensive lifestyles in a world in which,
for them, time and space have ceased to have much meaning. They
use air travel, telephones, faxes and computer networking to alternate
regularly and repeatedly between interest communities at different
scales up to the world-wide. For them, participation is everything;
place is incidental. Such individuals approximate the archetypal citizen of the global village and may be pointers to the ways in which
we shall all live our lives at some distant time in the future. Even in
the developed economies, where role specialisation and technology are
highly advanced, they are, however, very few in number. Such active

120

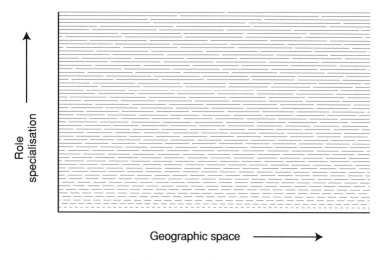

Figure 21   Webber's concept of role specialisation and its
geographical relationships

technology-based involvement is negligible in developing countries.
The power of place may be in decline but it remains strongly
entrenched for the majority of people over large parts of the world.

Passive citizenship of the global village is, however, substantial,
widespread and increasing rapidly. An argument can be made that
a global urban society is being created in the form of a world-wide
community of radio listeners, and television and video-cassette
viewers who see and interact with the city as an image, not a place.
Their urban experiences are predominantly electronic. Most broadcasting takes place from and is about cities, so this armchair audience
receives a diet of programmes and material which projects and in
many cases endorses and glamorises urban ways of living. Viewers
are exposed to pictures of cities as places, and the activities and
dialogue about the lifestyles and patterns of behaviour which take
place within them. Urban living is promoted by films and broadcasts which endorse capitalist consumption values. Urban fashions
and fads are publicised and the relationships which underpin urban
communities are explored and popularised. Fictional characters like
J. R. Ewing and Rambo become urban role models, and action,
violence and extremism, which are visually exciting, are emphasised.
The global impinges ever more on the local as viewing times and

audiences increase (Janelle, 1991). Traditional values are in retreat as mass media urbanism progressively permeates the most distant and remote rural societies.

The extent to which such imagery is extending urbanism across the globe, and, if it is, what form that urbanism takes, is a matter for debate and dispute. Global urbanism is a simple, superficially plausible but highly questionable concept. Three key questions are involved. The first concerns the size and character of the global market and addresses the extent to which media products are available to and are consumed by world-wide audiences. The second surrounds the urban content and significance of this material, while the third is about the impact of this content on the behaviour and ways of life of listeners and viewers. Such questions cannot be answered definitively, due to the limited research on international communication and its impacts. Two crucial deficiencies occur in work on the flow of communication between countries and on the cultural biases in those flows (Pool, 1976). Both have been written about extensively but there have been few empirical studies, and by far the greater bulk of that literature consists of polemical essays unenlightened by facts. Against this background the following sections attempt to synthesise the largely speculative views that have been advanced concerning the consequences for urbanism of the growth of global culture.

MEDIA SERVICES

Media urbanism has its roots in the newspaper and film industries of the late nineteenth century, but is essentially a creation of modern telecommunications. It traces its origins to the introduction of the first commercial radio broadcasts in Pittsburgh in 1926 and public television service by the BBC in London in 1936. Together, these developments initiated a process through which the world has become progressively interlinked for instantaneous communication. Today, commercial radio services are ubiquitous and some 150 of the 189 sovereign states and territories surveyed by UNESCO in its World Communication Report (1989) operate a television service. The principal countries which do not have their own television service are in Africa, and include Botswana, Cameroon, Central African Republic, Chad, Gambia, Malawi, Rwanda and Western Sahara. Several of these are very small countries which share the television service of a larger neighbour.

Global telecommunications were made possible by the launch in 1965 of INTELSAT (Early Bird), the first commercial satellite. Early Bird doubled the telephone capacity between the United States and Europe and for the first time enabled live television transmissions between the two continents to occur. Today there are over 160 communications satellites in orbit and global television is a reality. The number of sets rose from 192 million in 1965 to 884 million in 1992. Some 500 million people watched the moon landing in 1969 and around a billion saw some part of the 1976 summer Olympics in Montreal. In 1984 perhaps two billion saw the Los Angeles Olympics. Over three billion probably watched the 1992 games in Barcelona.

Although the technology was developed over seventy years ago, the time-lags involved in establishing networks of transmitting stations, developing programming schedules and in the spread in the ownership of receivers meant that, in many countries, television only began to reach a mass audience in the last twenty years. In others, the level of market penetration remains negligible. The USA and the United Kingdom, where television became a mass medium as early as 1953 following the successful broadcast of the coronation of Queen Elizabeth II, are exceptions. More typical is the case of China, where early experiments with television did not begin until 1956 and it was not until 1978 that the China Central Television organisation, which coordinates broadcasting over most of the country, was established. Television in China did not begin to enter the homes of working people until the 1980s (Lull and Se-Wen, 1988). Venezuela is similar in that television did not become a national service until the late 1960s (Barrios, 1988).

The effect of variable introduction and the initial high costs of installation both of broadcasting stations and receivers in the home is that the size of television audiences varies widely. Most of the television receivers presently in use are in the developed world and around one quarter of the total are in the United States. Among the ten countries with the largest number of receivers, only China, the world's most populous country, and Brazil are in the developing world. There are more television receivers in use in the United Kingdom than in the whole of Africa, and more in Japan than in South America (UN, 1994). The extent to which television is presently a communication medium of the developed world is underlined forcefully when allowance is made for population (Figure 22). The number of televisions per 1000 inhabitants is high in North America, Japan, Australia and parts of Europe, the Middle East and

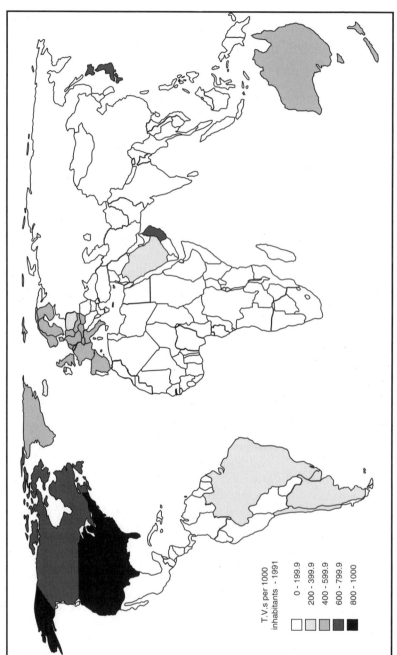

T.V.s per 1000
inhabitants - 1991

0 - 199.9
200 - 399.9
400 - 599.9
600 - 799.9
800 - 1000

Figure 22 Televisions per 1000 inhabitants, 1991

South America, but is very low throughout most of Africa and Asia. Among the major countries for which data are available, there are fewer than five televisions per 1000 inhabitants in the Central African Republic, Chad, Ethiopia, Mali, Mozambique, Tanzania, Zaire and Myanmar. In these and other developing countries, the lack of electricity services in rural areas is a major impediment to the spread of television.

It is important to emphasise that a simple mapping of the number of sets per head of population provides a very crude measure of the local importance of television. Figure 22 overstates the position in the United States and Japan since, in houses in which there is a set in each room, most are used infrequently. Conversely, in countries in the developing world, television is watched by many more people than the number of receivers suggests. Multiplier effects are high because even a small set can be viewed by several people at a time. With a newspaper, even two people cannot read a single copy comfortably. The importance of the extended family in traditional societies makes a home with a television a place where relatives and neighbours gather to watch, and so serves to extend any impacts. In the major towns almost every social or sports club has a colour television set. Television enjoys considerable novelty appeal in many parts of the world and is the principal means of entertaining and informing in societies in which the majority of the population is illiterate. It is more influential than radio because the information content of pictures is far greater than that of sound. What is transmitted, good or bad, reaches billions of viewers irrespective of age, colour, sex, religion or level of literacy.

URBAN IMAGERY AND VALUES

Television tends to be rich in urban content and imagery because most broadcasting companies are city-based and so direct their attention towards civic affairs and happenings. Cities are places in which the major political events take place so they are the focus of attention in most news and current affairs broadcasting. So effective are networks of information-gathering and dissemination that news bulletins around the world tend to carry the same stories, so emphasising the importance of the principal world cities. The extremes of urban poverty and affluence are a common focus of inquiry in documentary programming, while the cultural importance of cities is projected through the broadcasting of theatre and opera which are essentially urban art forms. The concentrated activity and action which takes place in cities is visually exciting and makes for supposedly 'good' television drama. Urban settings form the backdrop to most soaps and telenovelas, and the values of competitiveness, conspicuous consumption, selfishness and infidelity, commonly portrayed in American series such as 'Dallas' and 'Dynasty', are readily taken to be urban values.

Irrespective of where they live, audiences around the world are fed a broadly similar diet of television. The same types of programmes are scheduled at the same times of the day. Broadcasting tends to follow the same basic pattern because it is geared to people's daily lives. Soap operas and quiz shows account for most of the daytime slots, while children's programmes predominate in the early evening. These are followed by family viewing, the mid-evening news, drama, sport and adult television. The significance of this standard format is that it generates demands for particular types of programming, much of which is international in origin.

Although most countries have their own television services, few are able to generate sufficient output to fill their schedules and so they rely heavily on imports to make up the difference. It is estimated that around one-third of total television programme time across the world is imported material. This figure, however, must be regarded as a very crude approximation as the surveys undertaken by Varis in 1973 and 1984 (UNESCO, 1989) and in 1986 by Berwanger (1987) are based upon data supplied by only half of the world's sovereign states. They are especially deficient in their coverage of Africa. The United States is the biggest importer of television programmes, but it is the smallest importer in relation to its total television programme output (Table 10). Many countries rely on imports for more than half of their programming and it seems likely that most of the African countries for which information is not available fall into this category.

An example of a country which makes significant use of imports is Venezuela, where it is estimated that some 51 per cent of television is foreign material, the majority from the USA. Television programming in Venezuela has some similarities in its composition with the major American networks, in that it is possible each week to watch major serials including 'Falcon Crest', 'Dynasty', 'Dallas', 'Miami Vice', 'Kojak' and 'He-Man'. Nearly half of daily programming is made up of American films and children's programmes. Many of the tele-images that Venezuelans receive at home are from Los

Table 10 Estimates of imported television in selected countries, 1986
(as a percentage of total programming)

Below 10%			
USA	2	India	8
Japan	6	USSR	8
China	8		
Between 11% and 30%			
Indonesia	12	Netherlands	25
Philippines	12	Hungary	26
Republic of Korea	12	Bulgaria	27
Pakistan	16	Vietnam	28
France	17	Belgium	29
United Kingdom	17	Yugoslavia	29
Italy	18	Ethiopia	30
Germany, Federal Republic of	20	German Democratic Republic	30
Australia	21	Norway	30
Cuba	24		
Czechoslovakia	24		
Between 31% and 50%			
Canada	32	Portugal	39
Syria	33	Turkey	39
Venezuela	33	Argentina	40
Mexico	34	Nigeria	40
Egypt	35	Sri Lanka	40
Sweden	35	Denmark	43
Finland	37	Austria	43
Kenya	37	Chile	44
Spain	37	Democratic Yemen	47
Uganda	38	Malaysia	48
Brazil	39	Ivory Coast	49
Greece	39		
Over 50%			
Senegal	51	United Arab Emirates	65
Algeria	55	Ecuador	66
Singapore	55	Iceland	66
Tunisia	55	Brunei	70
Ireland	57	Peru	70
Mauritius	60	Zaire	70
Cyprus	60	New Zealand	73
Zimbabwe	61		

Source: UNESCO (1989: p. 148)

Angeles's streets, Miami's beaches, Dallas's offices, Manhattan's avenues and thousands of suburban houses of American middle-class families (Barrios, 1988).

The demand for imports is high because of the economics of contemporary broadcasting. Television is a Western invention which has spread slowly and unevenly throughout the world over the last half century. Much of the equipment, most of the know-how and many of the programmes still originate from Europe and North America. This dominance may be seen as a form of neo-imperialism, but the reasons are financial rather than overtly political. Low-income countries with embryonic television services have little choice but to buy programmes from abroad. In most countries the basic decision to broadcast even five hours per day is enough to create the need to import most of the material. Few developing countries can afford to commit the resources to produce 2000 hours of programmes of reasonable quality per year. Indeed the money for production actually available to individual television stations is often barely sufficient to produce the most rudimentary talk shows, amateur drama or variety programmes. Even if subtitles or dubbing are required, the low price of many imported programmes and the lack of local suppliers mean that station managers have to rely heavily on imports.

Statistics assembled by UNESCO (1989) emphasise the extent to which producers in a small number of Western countries dominate the global media market (Table 11). Some thirteen of the twenty largest media corporations are American and the rest are Japanese or European. Of the seventy-eight firms listed in the complete UNESCO table, forty-eight are American or Japanese, and not one is based in the Third World. Media moguls such as Rupert Murdoch, Silvio Berlusconi, Ted Turner and Henry Luce with Warner Brothers control corporate empires which span continents and include holdings in broadcast, print and film production, and also distribution facilities such as television stations, satellites and cable networks. As an example, the merger in March 1989 between Henry Luce and Harry and Jack Warner made Times-Warner the largest media corporation in the world. It has an estimated value of US$18 billion, employs some 340,000 people and operates from corporate bases in the United States and through subsidiaries in Australia, Asia, Europe and Latin America (Sreberny-Mohammadi, 1991). Time-Warner's own publicity materials describe the corporation as 'a vertically integrated global entity' and claim that 'the world is our audience'. Sony's corporate ethos is similarly one of 'thinking globally but acting locally'.

Producers from the USA, the United Kingdom, Canada and Australasia have a competitive advantage in the global television market because they work predominantly in the English language

Table 11 Major information and communication groupings, 1989

Group	Country	Ranking-media	Media sales US$ mn.	Press, publishing, recording (%)	Radio-TV, motion pictures (%)	Period
Capital Cities/ABC	USA	1	4,440	23	77	
Time	USA	2	4,193	61	39	
Bertelsmann	Fed Rep of Germany					June 87
News Corp	Australia	3	3,689	54	18	June 88
Warner Communications	USA	4	3,453	58	32	
General Electric	USA	5	3,404	49	51	
Gannett	USA	6	3,165		25	
Times Mirror	USA	7	3,079	88	12	
Gulf + Western	USA	8	2,994	85	11	
Yomiuri Group	Japan	9	2,904	37	63	
CBS	USA	10	2,848	63	23	86
ARD	Fed Rep of Germany	11	2,762		100	
NHK	Japan	12	2,614		100	
Advance Publications	USA	13	2,541		100	March 88
MCA	USA	14	2,397	92	8	
Knight Rider	USA	15	2,052	8	92	
Tribune	USA	16	1,973	90	5	
Asahi Group	USA	17	1,961	68	22	
Hearst	Japan	18	1,840	69	17	85
Fuji-Sankei	USA	19	1,835	79	16	
	Japan	20	1,825	59	41	81

Source: UNESCO (1989: p. 104)

(De Stefano, 1990). English is the main international language not because it is the first language of most people but because it is their second, third or fourth language. Wardhaugh (1987: p. 35) estimates that 'about one fifth of the world's population of better than five billion people has reason to use English almost every day'. These figures still put English behind Mandarin Chinese in numbers, but Chinese is not an international language – it is largely restricted to the mainland of China, where it is the language mainly of native speakers. Alternatively, if the estimate of Stevens (1987; p. 56) of 1.5 billion users of English is accepted, then it is ahead of Chinese.

Other languages such as Spanish, Hindi, Arabic and Portuguese may soon surpass English in terms of numbers of native speakers, but none so far has the international status of English. It is the single official language in twenty-five countries and a co-official language in seventeen more. Its nearest competitor is French, which is official in nineteen countries and co-official in nine (Wardhaugh, 1987: p. 135). English is institutionalised as the language of government and education in many countries in which there is no common indigenous language. It is the primary linking language in several countries such as India where there are many local or regional languages. Although only 3 per cent of the population is actually bilingual in English and Indian languages, much of government and higher education, and approximately half of all the books published in India, are in English. Global pre-eminence is reinforced because English is the main language of science and technology. It is the language for which most computer software is written.

MEDIA URBANISM

Despite the existence of a large and rapidly expanding global audience for media products, and the domination of supply by Western English-language producers, the implications for the spread of urbanism are much disputed. The arguments surround the power and effectiveness of television as an agency of social transformation and the directions which such media-induced changes are taking. Two alternative perspectives can be identified in the recent literature. The more traditional sees television as imposing a uniform urbanism of a Western variety upon and to the detriment of indigenous cultures. The more recent highlights the complexity of television flows and, as countries increasingly draw their programmes from a wide variety of sources, argues that the effects will be many and varied.

The traditional view is that mass media are creating a homogeneous urban culture which is being spread to all parts of the globe by television and related technologies. This culture is heavily infused with Western urban imagery which, as it washes across the world, is progressively displacing residual ruralism (Schiller, 1976; Mattelart, 1979). As such it can be conceptualised as 'cultural imperialism', in the sense that media dominance is seen as the logical successor to economic and political hegemony. This model rests upon two assumptions. The first is that television has the power to override all the other traditions and institutions that shape local society. The second is the supposed ubiquity and popularity of Western and especially American television and the belief that it will displace local media forms, to the detriment of parochial cultures. Despite the highly questionable bases for these assumptions, 'it is a thesis which has been spoken about in countless conferences, seminars, books and pamphlets and mouthed so loudly that it has been transferred from postulate to certain truth' (Tracey, 1993: p. 164). It has alarmed politicians such as the French Minister of Culture who in 1982 identified 'Dallas' as a national threat and called for a crusade against financial and intellectual imperialism that 'no longer grabs territory, or rarely, but grabs consciousness, ways of thinking, ways of living' (*ibid.*: p. 178). The long-term implication is that the world will be subsumed by a homogeneous urbanism in which everyone everywhere will live a way of life dominated by urban values.

Recent research, however, suggests that mass media are in fact contributing to the creation of highly varied patterns of urbanism at the global scale, and that the ideas of cultural imperialism lack empirical support (Sreberny-Mohammadi, 1991; Tracey, 1993). The concept of a plural urban culture is grounded in analyses of the international market in media materials which highlight the complex pattern of production, content and distribution. The United States and other Western nations are no longer the dominant producers of media materials. Moreover, their products do not always have a dominant presence in the countries into which they are imported, nor do they necessarily attract large audiences. Where there is a choice, domestic programmes are invariably preferred. Local output is not threatened by imports and in most countries is rising rapidly. It may project images of life in the city at the expense of those in the countryside, but such urbanism is extended predominantly within the context of national rather than global culture.

There is growing evidence to support the plural urban culture model. The most basic is the sheer increase in the volume of media material being generated in the developing world. Mexico and Brazil are major producers and distributors of television programmes, with TV Globo, the Brazilian network, exporting telenovelas to 128 countries including Cuba, China and the former USSR (Tracey, 1988). The Indian film industry is the most productive in the world, making nearly nine hundred films in 1985 (Dissanayeke, 1988). Most are Hindu epics in which traditional cultural themes and values are explored and reaffirmed. Egypt is an emerging centre of film-making in the Arab world.

Increasingly complex flows and exchanges of television programmes are revealed in studies of the global television market which were undertaken for UNESCO by Varis (1984; 1993). These challenge the established view that Western urban values are spreading across the world via a strong one-way flow of television programmes from the North Atlantic basin to developing countries. For example, in Arab countries, the Varis studies show that 42 per cent of television is imported, of which one-third comes from other Arab states. The United Arab Emirates, Egypt, Saudia Arabia and Kuwait are the main suppliers. France provides about 13 per cent and the United Kingdom, Japan and Germany between 5 and 7 per cent each. Similarly, in Africa, about 40 per cent of programmes are imported, though again there are wide differences in volume between individual countries. Some 50 per cent comes from the United States, 25 per cent from Germany and the rest is principally from Western Europe. In South Africa, 30 per cent of programmes are imported: 54 per cent from the United States, 30 per cent from the United Kingdom, 9 per cent from France, 5 per cent from Austria and 3 per cent from Canada. In Europe, according to Varis, more than 40 per cent of imported programmes originate within other countries in the region itself.

These, and similar data from other parts of the world, point to the existence of highly intricate patterns of media production and exchange. The number of producers is increasing and regional markets operate and are growing strongly. Close links exist between suppliers and consumers within particular language groups so that elements of cultural distinctiveness are maintained. Portugal, for example, is a major importer of Brazilian television programmes. Flows of videocassettes are broadly similar to those of television (Alvarado, 1990). The United States may have dominated the market when film and television were in their infancy, but the contemporary situation is one of multilateral production and trade in media products.

A second argument is that imported material is rarely as popular as domestic programmes, and is frequently used to fill the off-peak slots. Thus in Japan, 'Dallas' was introduced in October 1981 and went to only a 3 per cent market share in December. It never remotely compared with the popularity of 'Oshin', a locally produced six-days-a-week fifteen-minute serial drama which regularly attracted over half of the national audience. Locally produced telenovelas are far more popular in South America than imported American soaps. For example, in the mid-1980s, TV Globo broadcast a telenovela called 'Roques Santeriro' which regularly captured 90 per cent of the national market. Similarly in Venezuela, while US and Brazilian soaps are popular, they do not compete for the prime-time spots filled by local productions such as 'Cristal' and 'Los Donatti' (Patterson, 1987).

The preference for domestic productions is strong even in the Netherlands, where there are many competing channels and a sophisticated and highly educated audience. Dutch viewers can receive television from several neighbouring countries including the United Kingdom, and as many of them can understand French, German and English, there is a wide range of choice. The work of Bekkers (1987), however, shows that even in this extreme case of a pan-European market, the preference for domestic channels is strong. The two leading Netherlands channels have an overwhelming market share, which increased during the 1980s in spite of competition from satellite broadcasting. For Tracey (1993), evidence of this kind confounds the suggestion that imported television is not only ubiquitous but is also immensely popular. Instead it serves to emphasise the continuing strength of national cultures and the power of language and tradition through which they are perpetuated.

The evidence reviewed by Tracey is not unequivocal but it identifies a different urban present and points to a different urban future to that suggested by cultural imperialists. The view of Lull (1995) is that despite the world-wide reach of technology, we do not and will not live in a global village where an all-encompassing and uniform Western urbanism replaces outdated and unwanted rural ways of life. Mass media are extending the reach of urban values, but they are not impacting upon local cultures in the same way. The interaction between global and local results in a range of hybrid cultural forms. Just as living in a city does not create a homogeneous urban lifestyle in a single society, as was emphasised in the last chapter, so the world-wide transmission of information produces richness and variety of urban responses at the global scale.

CONCLUSION

The growth and spread of urbanism amounts to social change on a vast scale. It originated in the industrial metropolis, where the concentration of people from diverse backgrounds gave rise to and enforced a variety of lifestyles that were fundamentally different to those which existed in rural areas. Such patterns of association and behaviour were initially seen as the products of social breakdown as traditional forms of social control, based upon the membership of primary groups, were destroyed by the size, density and heterogeneity of urban populations. Later research suggested that they were the product of constrained choice based upon class, ethnicity and stage in life cycle. The rich variety of lifestyles is viewed in more positive terms by stressing the contribution made by size, critical mass and interaction to the creation of a mosaic of distinct subcultures in the city.

Whatever its causes, urbanism was initially restricted to cities from where it diffused locally by direct contact to encompass many people who lived adjacent to the city boundary. Urban values today, however, are spreading across the globe in the form of images and values carried by the mass media. Large parts of the world are presently permeated by urban values, and urbanism touches and impinges upon traditional value-systems in many of the most remote and traditional rural localities. World-wide social relations have intensified so that distant areas are linked in such a way that local happenings are shaped by events occurring many miles away, and vice versa (Giddens, 1990). The extension of urban culture is likely to increase as the numbers and proportion of the world's population living in urban places, and the power of the media to spread urban values, rise.

The global spread of urban values was traditionally viewed by many in largely negative terms. At best it was equated with undesirable modernism. At worst it was seen within the framework of 'cultural imperialism', alongside the rise of the transnational corporation, as a means of maintaining and extending the pre-eminence of Western capitalism. A central assumption in this approach was that Western, and especially American, urban values, such as competitiveness, consumerism and individualism which were expressed explicitly in a variety of media genre as well as indirectly through advertising, were being exported to and so were devaluing the rich and ancient cultures of many Third World countries. Fears of 'cultural homogenisation' were voiced and arguments were made for 'cultural dissociation' from the global capitalist system as a means of ensuring autonomous development and the protection of indigenous cultures (Sreberny-Mohammadi, 1991).

Today, it is possible to see the spread of urban values across the globe more positively. The international media market is far larger and more complex than that suggested by the cultural imperialists. Western media domination has given way to multiple suppliers and complex cross-flows of media products. More Third World nations are producing and exporting media materials so the variety of urban imagery in the world market-place is increased. Rural values and ways of life are under attack, but from a variety of directions, both domestic and foreign, and in a number of complex ways. The picture is one of increasing urbanism, but not necessarily greater social uniformity. Rather than a single global village with a homogeneous culture, the trend is towards a plural world-wide urban society.

RECOMMENDED READING

Bird, J., Curtis, B., Putnam, L., Robertson, G. and Tickner, L. (1993) *Mapping the Futures: Local Cultures; Global Change*, London: Routledge.

A set of papers, drawn from a number of different disciplines, which assesses the links between local culture and global change.

Brunn, S. D. and Leinbach, T. R. (1991) *Collapsing Space and Time*, London: Harper Collins.

A collection of papers on the human consequences of a shrinking world, including useful contributions from Gold on electronic villages and Janelle on the implications of increasing global interdependence.

Curran, J. and Gurevitch, M. (1992) *Mass Media and Society*, London: Arnold.

A collection of review papers on media production and impacts. It includes an especially useful contribution from Sreberny-Mohammadi on global–local relationships in international communication.

Griswold, W. (1994) *Cultures and Societies in a Changing World*, London: Pine Forge.

A sociological examination of the complex interplay between culture and social structure using examples from Nigeria, China and the United States.

Noam, E. M. and Millonzi, J. C. (1993) *The International Market in Film and Television Programs*, Norwood NJ: Ablex.

A collection of papers on several aspects of media production and distribution. It includes especially useful contributions from Varis on trends in the global traffic of television programmes and Tracey on popular culture and the economics of global television.

Webber, M. M., Dyckham, J. W., Foley, D. L., Guttenburg, A. Z., Wheaton, W. L. C. and Wurster, C. B. (1964) *Explorations into Urban Structure*, Philadelphia PA: University of Pennsylvania Press.

An early but still relevant consideration of the relationship between interaction and spatial structure. Webber's chapter on the urban place and the non-place urban realm is especially stimulating.

7

WORLD CITIES

The urban world is dominated by a small number of centres that are the command and control points for global capitalism, the world's dominant economic system. Such centres are distinguished not by their size or their status as capital cities of large countries, but by the range and extent of their economic power. They are the locations for the key individuals, institutions and organisations which manage, manipulate, dictate and determine the formation and reproduction of capitalism across the world. These attributes give such cities a disproportionate and exceptional importance so that they occupy dominant positions in the global urban hierarchy. So preeminent is their status and so powerful and pervasive are their influences that they merit designation as world cities.

The concept of world cities, first coined by Geddes in 1915, was reintroduced and expanded by Hall in 1966. In his book *The World Cities*, he identified a set of places in which 'a quite disproportionate part of the world's most important business is conducted' (p. 7). Hall distinguished world cities from other places of great population and wealth because they were major centres of political power, seats of national and international government and concentrations of related professional bodies, trade unions, employers' federations and corporate concerns. They were also centres of trade, finance and communication. Characteristically they were great ports, distributing imported goods to all parts of their countries and in return receiving commodities for export to the other nations of the world. Within each country they were the focus of road and rail networks and the sites of major international airports. Such activities gave the cities distinctive social and economic characteristics which were reflected in their status as centres of professional talent in the fields of medicine, higher education, research, culture and the arts. They were known for their universities, hospitals, concert halls and museums. On these bases, the places recognised by Hall as being world cities in 1966 were London, Paris, Randstad, Rhine-Ruhr, Moscow, New York and Tokyo.

Over the last thirty years, the meaning of the term and hence the list of centres has been refined and revised as world cities have become a major focus of research interest (Friedmann, 1986; King, 1989; Sassen, 1991, 1994). Global economic status and influence have replaced the volume of business transacted as the key diagnostic criteria. Most observers today see world cities as the decision-making points for the world-economy. They are distinguished by their roles as sites for the accumulation and concentration of capital and as places from which its distribution and circulation are organised and controlled. As such they are the 'key nodes' of the international urban system (Mollenkopf, 1993). World cities are favoured locations for the institutions of international production and consumption and the individuals and agencies which support and facilitate these activities. Function rather than size is the key criterion. World cities are places in and from which global business, finance, trade and government are orchestrated and arranged.

The command and control role is reflected in the activities which are typically located in world cities. These include corporate management, banking, finance, legal services, accounting, technical consulting, telecommunications, computing, international transportation, research and higher education (Friedmann and Wolff, 1982: p. 320). Concentrations of such functions are common in capital cities which serve as the highest-order service centres for their national urban systems. They are especially pronounced in the principal business centres of the richest and most advanced economies. The number and range of services do not of themselves, however, denote world status since outlook and orientation are key considerations. What distinguishes world cities is that they provide services for a world market and not merely for domestic consumption.

World cities are characterised by concentrations of headquarters of global corporations; commodity, currency and securities exchanges; and head offices of producer services organisations. They are major centres for international government and administration and are principal junctions on the global conference and convention circuit. The importance and distinctiveness of these activities are conveyed visually by office and convention buildings of distinctive architectural design (Zukin, 1992). The presence of such functions and institutions means that world cities have more in common with each other than they have with urban centres in their own countries or with places of similar size elsewhere. The strength of their inter-linkage means that it is often as easy to travel among them as it is to reach secondary centres in the same country or in adjacent terri-tories. World cities are the principal foci for global business travel and telecommunications networks.

The presence of large numbers of members of the transnational corporate and producer service class means that world cities have a distinctive sociology which is expressed in terms of occupation, income and ways of life of their residents. They are places of social polarisation which may be conceptualised as 'dual cities' (Castells, 1989). Those who work in the organisations and institutions which sustain world city functions constitute a well-educated, socially mobile, footloose and highly paid elite. They are cosmopolitan in origin and global in outlook. Their corporate, diplomatic and profes-sional skills are well developed, highly prized and generously rewarded. The interests, attitudes and aspirations of these employees typically define and maintain a lifestyle and culture which further distinguish global cities from other principal business centres and which are far removed from those which are experienced at the oppo-site end of the social spectrum. A low-skilled and low-paid working class which services the international service sector exists alongside but well separated from the community of global professionals. It typically includes large numbers from ethnic minority backgrounds. World cities are places of exceptional wealth and affluence, but they are also places of severe disadvantage and deprivation.

The cities which are recognised as being of global importance today vary according to the mix of diagnostic criteria used. Friedmann's list in 1986 consisted of thirty centres arranged in a 'world urban hier-archy' (Table 12). These include primary and secondary cities in the core countries with industrial market economies, and primary and secondary centres in semi-peripheral countries with upper middle income market economies. His criteria were: status as a financial, manufacturing and transportation centre; number of headquarters of transnational corporations; number of international institutions pre-sent; rate of growth of business services; and population size. Thrift (1989) recognised three main strata of world cities. At the top there are truly global centres which contain many head offices, branch offices and regional headquarters offices of the large corporations and banks (Table 12). They account for most of the major business

Table 12 Hierarchies of world cities

Core countries		Semi-periphery countries	
Primary	*Secondary*	*Primary*	*Secondary*
Europe:			
London	Brussels		
Paris	Milan		
Rotterdam	Vienna		
Frankfurt	Madrid		
Zurich			
America:			
New York	Toronto	São Paulo	Buenos Aires
Chicago	Miami		Rio de Janerio
Los Angeles	Houston		Caracas
	San Francisco		Mexico City
Asia:			
Tokyo	Sydney	Singapore	Hong Kong
			Taipei
			Manila
			Bangkok
			Seoul
Africa:			
			Johannesburg

Source:: Friedmann (1986)

Global centres	Zonal centres	Regional centres
New York	Paris	Sydney
London	Singapore	Chicago
Tokyo	Hong Kong	Dallas
	Los Angeles	Miami
		Honolulu
		San Francisco

Source:: Thrift (1989)

dealings on a world scale. Second, there are zonal centres which have corporate offices of various types and serve as important links in the international business system but are responsible for particular zones rather than for business on a world scale. Third, there are regional centres which host many corporate offices and foreign financial outlets but are not essential links in the international business system. The gaps between these strata are wide and there are fundamental differences in the global role and importance of cities in the first two tiers. Despite variations in the composition of lists, most analysts point to the pre-eminence of New York, London and Tokyo. London is no longer a mega-city, New York is neither a national nor a state capital, but by virtue of their concentration of globally important institutions and organisations these three cities are the principal command and control points for international business and commerce (Sassen, 1991).

Despite structural and functional similarities, world cities have particular product and service specialisms which reflect their history, location and the size and character of their national economies. In this respect they are complementary rather than competing. New York, reflecting the manufacturing strength of the United States' economy, is the principal locus of global corporate power. It is the main centre of global political power and a place from which a sizeable component of global production and consumption is controlled. Tokyo is a world city principally because of the recent success of the Japanese economy. Its status is largely self-generated because it is physically remote from the Western countries that are the traditional centres of the world economy, and because it was closed to the influx of foreign investment and immigration for many years. As the main export point for national financial wealth it has a strong orientation towards its domestic sources of money supply, as well as to world financial markets (Machimura, 1992). The manufacturing economy of the United Kingdom is comparatively small but London is a world city because it is the principal supplier of financial and producer services to global markets, a role which it developed as the hub of the British Empire (King, 1990). It is also the cultural centre of the English-speaking world. Though less important as a corporate and financial centre, Paris is the most popular location for the headquarters of international organisations and for international conventions, a position which owes much to its many architectural splendours and gastronomic attractions (Knight and Gappert, 1989).

Although the concentration of capitalist functions in a small number of cities is not in dispute, many observers question the extent to which a handful of cities can and do perform a world-wide role. The concept of world cities is attractive but the empirical substance is disputed. One argument, as discussed in Chapter 2, is that the global economy is itself a misnomer as so many countries and areas lie beyond its reach. For example, urban systems in large parts of Africa are presently embryonic and although their primate cities may be well connected, there are vast territories housing millions of people who exist beyond or at best on the extreme margins of the global urban economy. Their world is highly localised and their world cities are the nearest market or distribution centre. A second point is that it is easy to overstate the scale and extent of the power which is discharged from world cities. Although the capitalist institutions in world cities exercise an element of global control, it is important to emphasise that they are themselves influenced and shaped by global politics and economics. At times they seem to be all-powerful, but there have been many world events, such as the Gulf War, which they were unable to prevent and which have had a powerful effect upon them. There is more to this debate than the validity and meaning of descriptive labels. It addresses the key issues of how far the contemporary world hangs together as an economic and an urban system and how far that system is dominated by and orchestrated from a small number of points of capitalist leverage.

The rise of world cities both reflects and has made possible the emergence of the world-economy. Whether their title is appropriate or misleading, they are the key locations in which global economic power is located and from where the world-economy is manipulated and controlled. Three factors are responsible for this locational and organisational pattern: the growth of the number and range of the institutions of global capital, their geographical concentration, and the recent extension of global reach via telecommunications and transport. It is in relation to these factors that the emergence and role of so-called world cities, exemplified by London, New York and Tokyo, can best be analysed and understood.

THE GROWTH AND ACTIVITIES OF THE INSTITUTIONS OF GLOBAL CAPITAL

World cities exist because they are the chosen locations for the agencies of global capitalism. They are places in which the principal business functions have developed and are concentrated and from which global corporate, financial and political control is exercised. A wide range of organisations is responsible for the accumulation and reproduction of global capitalism including those concerned with goods production, financing, and the provision of advanced producer and personal services (Table 13). It is the combined presence of such activities and the global roles which they perform that distinguishes world cities from mere national or regional centres.

Table 13 Activities and organisations of global capitalism

Commodity	Activities	Market structure	Organisations
Manufactured goods	Industrial production	World markets	Transnational corporations
Money	Borrowing, lending	Wholesale money markets, foreign exchange markets, Eurocurrency markets	Banks, discount houses, foreign exchange dealers
Financial securities	Securing of debt, speculation	Primary and secondary bond markets, financial futures markets	Banks and securities houses
Currency	Change, speculation, risk avoidance	Foreign exchange markets, Eurocurrency markets	Banks, foreign exchange dealers
Stocks and shares	Issuing, broking	Stock exchanges	Issuing houses, stockbrokers
Raw materials	Merchanting, broking	Commodity markets, futures markets	Brokers, merchant banks
Insurance	Underwriting	Insurance markets	Lloyd's and other insurance organisations
Freight	Chartering	Shipping exchanges	Chartering companies
Accountancy, legal, tax, advertising, public relations, management consultancy services, etc.	Provision of professional services	World markets	Companies
Transport	Ticketing, carriage	World markets	Travel agencies, airlines, rail companies, car rental companies
Hotel	Provision of food and accommodation	World markets	Hotels and restaurants
Personal finance	Provision of cash, currency and credit	World markets	Banks, credit and charge card companies

Transnational corporations are the most important agencies and their proliferation and increase in size are both causes and consequences of the emergence of the world-economy, as Chapter 4 has shown. The task of administering and managing global production requires elaborate corporate bureaucracies which are typically organised in a hierarchy of offices. At the apex is the global head office, normally in the country of origin, where strategic decision-making takes place and from where corporate empires are coordinated and controlled. It is here that the long-term future of the corporation is shaped and where operating structures and production targets for the constituent parts of the corporation are determined. Implementation of such corporate policies is typically the responsibility of regional head offices in each country or territory in which there is a significant production or marketing presence. An important role for staff at regional head office is to integrate the corporation into the local business culture so as to minimise the potential sensitivities which surround external ownership and the remission of profits. Tactical decisions concerning day-to-day operations are typically taken at plant level by managers whose responsibility it is to organise and maintain high levels of profitable production. Their job involves managing the receipt of raw materials and component parts, organising production runs, and arranging for the dispatch of finished goods to the market. This hierarchical structure is exemplified by the transnational farm machinery manufacturer Massey-Ferguson, which has its head office in Toronto, a European regional office in Coventry UK, and production control offices at each of its principal manufacturing plants in North America, Europe and Asia.

The growth in international production has been made possible by the development of an elaborate system of global finance which is organised through a number of different markets, mediated and controlled by a wide variety of finance capital institutions (Table 13). Complexity is the overriding characteristic and the distinctions between the various types of financial commodities and the markets and sub-markets in which they are traded are imprecise and blurred. One firm's investments are another firm's borrowings, and capital appears in a number of different forms, including cash, equities, securities, bonds, futures, bills of exchange and promissory notes, as it circulates around the system. Some markets are highly regulated while others are informal or 'grey'. They may have a physical identity in the form of a stock exchange building to which trading is restricted, or they may exist as a set of individuals, working for institutions, who communicate with each other so as to buy and sell financial commodities or products.

Banking, in which arrangements are made to attract money from investors and to make loans to individuals, corporations, institutions and governments, lies at the traditional heart of the global financial system. Merchant banks and exchange houses play a central role in the circulation of money, making profits for their shareholders by charging fees for their services. Both the size and the complexity of this financial operation have increased in recent years as the overall demand for funds for industry and economic development has risen, and as the sources of borrowing and destinations of investments have proliferated and diversified across the globe. Traditional forms of lending and borrowing through banks, however, are declining in relative importance as alternative financial products, markets and institutions are being introduced. The Third World debt crisis of the early 1980s, in which some thirty-five sovereign borrowers were unable to service their debts, was the turning point. In response, many countries switched into bonds as a means of raising and servicing finance (Daniels, 1993; Sassen, 1994).

The buying and selling of the currencies which are required for transactions in international goods and services takes place in foreign exchange markets. Most foreign exchange transactions involve the buying or selling of US dollars which function as an unofficial world currency. The fact that the value of currencies is liable to fluctuate, bringing gains or losses to those who hold it at the time, means that foreign exchange has become an important commodity in which institutions speculate. Substantial trade takes place in currencies as dealers and brokers seek to make profits for their clients, and fees for themselves, through judicious buying and selling. Such currency trading has been helped by the recent internationalisation of domestic money, so that most of the principal currencies are traded across the world and around the clock. For example, the domestic currencies of the United States and Japan are freely traded in Europe as Eurodollars and Euroyens respectively in the rapidly growing Eurocurrency market. Currency trading has increased in size and complexity in recent years through the development and exchange of swaps through the so-called swaps market. Swaps are mutually beneficial arrangements between two parties to alter interest rates or currency exposure, so as to avoid the need for unnecessary and perhaps unfavourable foreign exchange dealing. A further refinement is the growth of trading in agreements to buy currencies at specified prices at specified dates, again as a means of minimising risks and ensuring supplies. Exchanges such as the London International Financial Futures Exchange (LIFFE) facilitate this activity.

A separate and expanding set of activities is concerned with the issuing, buying and selling of financial securities. These are redeemable at a specified time and so introduce an element of stability into what could otherwise be a dangerously volatile finance market. Financial securities typically take the form of bonds or promissory notes which are issued by borrowers and give the lender a specified income at an agreed maturity date and interest payments at regular intervals before that date. There are also numerous variants on fixed-rate bonds together with other methods of securing debt obligations such as junk bonds. Banks manage the issue of new bonds and also trade in the so-called secondary market in which bonds that have already been issued are bought and sold. The value of bonds issued in the first half of 1986 was US$57 bn. and an estimated 39 per cent of issues were handled by just five leading managers: Credit Suisse First Boston (US/Switzerland), Deutsche Bank (Germany), Merrill Lynch (US), Salomon Bros (US), and the Japanese firm Nomura Securities (Daniels, 1993).

A fourth set of global financial activities is concerned with stocks and shares. These securities are issued by brokers on behalf of governments and companies, and are sold on the primary market to raise capital. They are then traded in a speculative fashion on the secondary market. Governments across the world apply strict regulations to the buying and selling of these items and dealing is restricted to stock and security exchanges. Although their traditional role was domestic, the leading exchanges increasingly issue and deal in stocks and shares for global companies and corporations.

A principal role of finance, however it is raised, is to expedite the buying and selling of industrial raw materials. A long-established set of markets and institutions is concerned with mediating the exchange of commodities, including agricultural produce (e.g. cocoa, coffee, sugar, rubber, wool, soya meal), metals (copper, lead, zinc, nickel, aluminium), bullion (gold, silver) and oil (Thrift, 1987). Some are sold for immediate delivery in the so-called spot markets, of which that in Rotterdam for oil is probably the best known. Most, however, are marketed via fixed-time, fixed-priced bills of exchange known as 'futures', which are used by corporations to ensure their supplies, to protect against commodity price rise and falls, or for speculation. They are traded in futures markets, examples of which include the

London Metals Exchange, the London Gold Futures Market and the Chicago Mercantile Exchange.

The growth of the international service economy is reflected in the rise of the advanced producer services sector which provides support services to industry. It includes the provision of professional services including insurance, accountancy, real estate, law, advertising, research and development, public relations, management consultancy, and office services such as stationery supply, cleaning and security. The change to global operation has been most marked in those service sectors in which the level of international activity was historically limited. One such field is accountancy, where the principal firms, based upon the number of audits of the largest 500 companies as identified in *Fortune* magazine, are sizeable multinationals (Thrift, 1989). The same applies to estate agencies such as Goddard and Smith; Gooch and Wagstaff; Healey and Baker; Hillier Parker; May and Roden; Jones Lang Wootton; Knight, Frank and Rutley; Pepper Angliss; Richard Ellis; St Quinton; Savills; and Weatherall, Green and Smith. In the last two decades, management consultants, property consultants, law firms, merchant bankers and advertising agencies have also expanded their operations into global markets (Leyshon, Thrift and Daniels, 1987a, 1987b; Moss, 1987; Perry, 1990). Such global business is facilitated by means of the organisation of employee services on an international basis. Examples include hotel accommodation (Hilton, Best Western, Holiday Inn), car hire (Hertz, Avis) and personal spending through credit and charge cards (Mastercharge, Visa, American Express).

THE CONCENTRATION OF COMMAND AND CONTROL

As manufacturing, financial and service organisations have increased in size and have extended their spheres of operation across the globe, they have concentrated their headquarters functions in a small number of cities which have grown in world stature as a result. Ten cities host the headquarters of nearly half (242) of the world's 500 largest transnational manufacturing corporations (Table 14). The top four cities alone account for the headquarters of 156. The remaining 344 headquarters offices are spread across forty-seven other cities. The New York metropolitan area is the principal choice, with fifty-nine headquarters (including eighteen of the top 100). London and Tokyo are of roughly the same importance, each being the

Table 14 Headquarters of the 500 largest transnational firms
(excluding banks) by city, 1984

City	Metropolitan area population (000s)	Number of firms
1 New York	17,082	59
2 London	11,100	37
3 Tokyo	26,200	34
4 Paris	9,650	26
5 Chicago	7,865	18
6 Essen	5,050	18
7 Osaka	15,900	15
8 Los Angeles	10,519	14
9 Houston	3,109	11
10 Pittsburgh	2,171	10
11 Hamburg	2,250	10
12 Dallas	3,232	9
13 St Louis	2,228	8
14 Detroit	4,315	7
15 Toronto	2,998	7
16 Frankfurt	1,880	7
17 Minneapolis	2,041	7
18 San Francisco	4,920	6
19 Rome	3,115	6
20 Stockholm	1,402	6

Source: extracted from Smith and Feagin (1987)

headquarters location of more than thirty transnationals. Seoul, with four, is the only Third World city which hosts the headquarters of firms in the top 500. The dominance of the top four cities is even greater when account is taken of turnover and sales. The central role played by the core countries in the world-economy is underlined by the fact that all the cities with ten or more headquarters are in the United States, the United Kingdom, France, Germany or Japan.

It is important to emphasise that it is the concentration of corporate control, and not population size, that distinguishes world cities. Tokyo and New York happen to have large populations and be centres of global corporate power, but six of the world's largest twenty metropolitan areas have no transnational headquarters (Table 15). Four others have only one. The lack of transnational headquarters in Beijing and Shanghai is understandable because China is only partially integrated within the world-economy, but otherwise the data emphasise the difference in the economic roles which are played by

148

Table 15 Headquarters of the 500 largest transnationals
(excluding banks) by size of city, 1984

City	Metropolitan area population (000s)	Number of firms
1 Tokyo	26,200	34
2 New York	17,082	59
3 Mexico City	14,600	1
4 Osaka	15,900	15
5 São Paulo	12,700	0
6 Seoul	11,200	4
7 London	11,100	37
8 Calcutta	11,100	0
9 Buenos Aires	10,700	1
10 Los Angeles	10,519	14
11 Bombay	9,950	1
12 Paris	9,650	26
13 Beijing	9,340	0
14 Rio de Janerio	9,200	1
15 Cairo	8,500	0
16 Shanghai	8,300	0
17 Chicago	7,865	18
18 Delhi	6,889	0
19 Philadelphia	5,254	2
20 Essen	5,050	18

Source: extracted from Smith and Feagin (1987)

the principal cities in developed and developing worlds. A handful of cities in the developed countries house all the top firms. Seventy-five per cent of the 162 cities in the world which have a population of over one million have no transnational firms' headquarters (Smith and Feagin, 1987).

Financial institutions are similarly concentrated in a small number of world cities. They give rise to an elite group of centres which transact a disproportionate share of the world's financial business. Reed's (1984) analysis of sixteen measures of financial status including capital, assets, deposits and earnings of international banks head-quartered in the centre; the volume of international currency clearings; the size of the Eurocurrency market; foreign financial assets and liabilities; the number of internationally active banks present; and indices of global linkage, suggests that this group has four levels (Table 16). The highest tier consists of supranational financial centres which are headquarters for a large number of internationally active banks that are well connected to other centres throughout the world.

Table 16 The hierarchy of global financial centres in 1980

Supranational centres, first order			
London	New York		
Supranational centres, second order			
Amsterdam	Frankfurt	Paris	Tokyo
Zurich			
International financial centres, first order			
Basel	Bombay	Brussels	Chicago
Dusseldorf	Hamburg	Hong Kong	Madrid
Melbourne	Mexico City	Rio de Janerio	Rome
San Francisco	São Paulo	Singapore	Sydney
Toronto	Vienna		
International financial centres, second order			
Bahrain	Buenos Aires	Kobe	Los Angeles
Luxembourg	Milan	Montreal	Osaka
Panama City	Seoul	Taipei	

Source: Reed (1984)

They are places where large amounts of foreign financial assets and liabilities are managed, where foreign direct investment capital is supplied to the rest of the world, and from where the organisational and operating norms that govern internationally active financial institutions are established. London and New York are the leading supranational financial centres on account of the size and diversified nature of their financial dealings. Amsterdam, Frankfurt, Paris, Tokyo and Zurich also perform a supranational role but have only limited global importance. The third level consists of eighteen international financial centres which host large numbers of foreign financial institutions and the headquarters of a small number of internationally active banks. They have only a limited capacity to influence events that affect global asset and liability management. These places play an important financial role within their regions but have only small stock markets and engage in relatively little truly international financial business. Host international financial centres occupy the lowest tier. The eleven cities assigned to this category have a relatively large number of foreign finance institutions from a large number of countries but are only small-time players in the global financial system.

More detailed analysis shows both the extent to which global finance is managed and controlled from a small number of world cities and the subtle differences of specialisation which exist among them. The degree to which the world's stock market activities are

Table 17 International stock market comparisons (as at 31.12.92)

Exchange	Market value (fixed interest and equities) £m.	Turnover £m.	Companies quoted domestic	overseas	total
New York	3,963,428	1,160,561	1,969	120	2,089
London	2,579,339	1,044,712	1,878	514	2,392
Tokyo	2,462,682	393,267	1,651	117	1,768
Osaka	2,406,181	95,682	1,163	5	1,168
Frankfurt	1,136,303	891,017	425	240	665
Luxembourg	937,188	736	59	162	221
Toronto	733,801	41,491	1,049	70	1,119
Paris	597,760	582,147	515	217	732
Milan	493,979	867,519	226	3	229
NASDAQ*	411,120	588,859	3,850	261	4,111
Brussels	287,493	6,834	164	154	318
Amsterdam	243,572	110,290	251	246	497
Sydney	208,224	32,915	1,038	35	1,073
Copenhagen	195,201	528,402	257	11	268
Montreal	178,449	11,487	556	22	578
Johannesburg	137,861	124,754	642	29	671
Switzerland	128,209	76,489	180	240	420
Korea	125,793	77,217	688	–	688
Madrid	113,286	27,095	401	3	404

* National Association of Securities Dealers Automated Quotations (a United States-based electronic market)

Source: The Stock Exchange, London.

concentrated in London and New York is remarkable and throws into sharp relief the pretensions of provincial and parochial centres such as Miami and Sydney to be world-ranking financial centres (Sassen, 1994). The global capital market is orchestrated by at most ten leading exchanges, with New York being by far the biggest in terms of market value (Table 17). With nearly £4bn. of fixed interest and equity stocks quoted it is significantly larger than London (£2.6bn.). Its size reflects the strength of the US economy and in particular the value of the many large transnational corporations whose stocks and shares are traded. The level of turnover and the number of companies quoted in New York, however, are almost exactly the same as in London. What distinguishes the London stock exchange from those in New York, Tokyo and Osaka, and under-lines its status as a global as opposed to merely a domestic financial centre, is the large number of overseas companies quoted. This is four times that in its principal rivals. A high level of international orientation is similarly a feature of the smaller exchanges in Frankfurt, Luxembourg, Paris, Amsterdam and Switzerland. The extent to which the major cities in the core economies command and control the global capital market is emphasised by Table 17. Of the twenty leading stock exchanges, only Johannesburg (ranked 17) and Korea (19) are in the Third World. Considerable growth has occurred in turnover at all stock exchanges in recent years but the rank order of the market hierarchy has remained relatively unchanged.

Tokyo and London are the world's leading banking centres, the former because it has the headquarters of the largest number of the major commercial banks, the latter because of its pre-eminence in international banking. Some seven of the world's fifteen largest banks, including Sumitomo, Sanwa and Fuji, which in March 1994 were the top three, are Japanese, of which five are based in Tokyo (Table 8). This geographical pattern was responsible, in 1986, for a greater concentration of bank assets in that city (US$1,804bn.) than in New York ($905bn.), Paris ($659bn.) and Osaka ($558bn.), as Noyelle (1989) observed. London, however, is by far the leading centre for foreign exchange dealing. For example, the daily average turnover in April 1989 was some US$190bn., a level that was higher than that in New York ($130bn.) and Tokyo ($110bn.) and far in excess of that in the other competing centres, of which Zurich ($55bn.) was the most important (Sassen, 1994). In 1994 London had 429 foreign banks while New York had 231 and Tokyo only 56. Tokyo in fact had fewer branches of foreign banks than Singapore (65) and Hong Kong (72).

All the top 100 banks in the world are represented in London and over 60 countries have a direct banking presence. Those from Japan are prominent and about 40 per cent of their international business world-wide is booked through their London branches or subsidiaries. British-owned banks and subsidiaries of foreign banks in London hold just under 25 per cent of all international claims booked in countries reporting to the Bank for International Settlements (Daniels, 1991). This market share is well in excess of that in New York (15 per cent) and Tokyo (10 per cent). In comparison with London, Tokyo is not as yet a world financial centre, its present status reflecting the strength of the Japanese economy rather than its importance for international banking and exchange.

The concentration of corporate and financial activities gives rise to and is supported by a related geography of producer service organisations. Firms which supply advanced professional services are concentrated in world cities where they support and in turn are supported by the major transnational corporations and finance houses. New York and London are the leading producers and exporters of international accountancy, advertising, management consultancy, legal and business services (Sassen, 1994).

The distinctive role played by world cities in the global economy is apparent when comparisons are drawn with principal domestic business centres. Lower-order places may provide a wide range of corporate and advanced services, but this is for local rather than world consumption. Such differences in orientation are especially marked in the United States, where all the major cities with the exception of New York function as regional service centres. Los Angeles is reckoned to be the second most important commercial city in the USA, but in 1985 it had only eleven banks where foreign exchange was traded, as opposed to 108 in New York (Levich and Walter, 1989). It had 111 foreign bank offices as compared to 405 in New York and seventy-nine in Chicago (Moss, 1987). Only one of the world's top fifty banks has its headquarters in Los Angeles (Noyelle, 1989).

BUSINESS INFORMATION AND DECISION-MAKING

The key individuals and institutions of international production, finance, services and government concentrate in world cities because these are the best places from which to direct global activities. A complex set of location factors is involved, including access to information, economies of scale, attractions of prestige locations, and exceptional global accessibility. Individual activities respond to these attractions in different ways. The fact that they support each other means that the benefits of world city location are cumulative and self-reinforcing.

World cities are the favoured locations for the headquarters of transnational corporations because they offer unparalleled access to business information. The principal responsibility of global corporate executives is 'orientation', which involves determining the nature and course of survival and growth of their organisations (Thorngren, 1970). This strategic task necessitates long-term scanning of socio-economic environments so as to identify and evaluate the factors that will affect the future operation of the business. The problem is then to choose a course of action that will minimise the threats to the firm and maximise the prospects for continued profitable operation. The aim of orientation is to ensure that the firm evolves in such a way as to be optimally placed in the future to take account of advantageous trading conditions. A wide range of influences needs to be considered, including raw material supply, the behaviour of markets, the activities of affiliates and competitors, developments in science and technology, the availability and cost of finance and labour, and government policies. Each is surrounded with uncertainty, since it is determined by external and contextual factors which lie outside and beyond the control of the firm.

Corporate orientation is a risky activity and the stakes are correspondingly high. It involves making informed choices on new, unpredictable and non-standard problems about which there are many unknowns. Effective decision-making requires good quality business intelligence about the future environment in which the firm will operate. Of particular value is information or informed opinion on events such as stock market trends, fluctuations in currency values, wars, political upheavals and changes of government that can influence and affect the level of global business. Intelligence of this type cannot be generated within an organisation, since it is to do with the behaviour of competitors, partners, politicians and states. It is most easily and reliably gleaned through regular face-to-face interaction with presidents, chief executives, directors and board members of global business, finance and producer services organisations. Members and officials of national and international governments who are similarly engaged in strategic planning and orientation for their own organisations also possess information which is of value in corporate orientation, as do research scientists who are engaged in work on new industrial processes and products.

Financiers are similarly drawn to world cities because of the access they afford to collaborators and competitors. The task of arranging loans, underwriting share issues and financing capital projects in foreign countries involves detailed and confidential negotiations with a wide range of government, banking and producer service organisations and it is of benefit collectively if the key representatives and decision-makers are concentrated in the same place. It is also important for purposes of prestige for foreign banks and finance houses to have a presence in the cities which are at the centre of world financial markets. Telecommunications could undermine the benefits of concentration but the evidence is that the principal financial centres are gaining rather than losing in importance. For example,

Table 18 Top executives in City of London financial institutions by educational background, 1986 (percentages)

| Institution | School | | University |
	Public*	Grammar**	Oxbridge
Merchant banks	83	8	72
Stockbrokers	96	4	80
Clearing banks	56	21	54
Accountants	72	18	52
Insurance companies	58	15	59
Insurance brokers	77	13	N/A
Foreign banks	Mostly foreign educated		51

Source: extracted from Bowen (1986: p. 39).

Notes * Independent schools at which attendance is on a fee-paying basis.
  ** Local authority controlled schools.

in December 1994 the Deutsche Bank announced that, because of the exceptional access to the market, it was to concentrate its investment banking business in London through Morgan Grenfell, the British merchant bank, rather than to push Frankfurt as the hub of its global corporate finance, share trading and derivatives operations.

Decision-makers benefit collectively from living and working in close geographical proximity and from the resulting opportunities for generating and accessing business information. Personal contact enables the characteristics of associates, partners and competitors to be scrutinised and assessed and contributes to the building of 'confidence', which is an essential prerequisite for successful business. It is helped by the fact that top executives are invariably drawn from the same narrow social backgrounds and have similar values and business ethics. Financial institutions in the City of London are especially inbred, the majority of the employees being public school and Oxbridge educated (Table 18). Men predominate in the top firms. The advantage of world city location is the exceptional access it gives to collaborators and competitors. Networking and corporate diplomacy take place at prearranged business meetings and conferences, and also when executives meet informally in clubs, restaurants and in hospitality suites at major sporting and cultural events. Corporate entertaining is an important medium through which business relationships are established and from which information, intelligence and contacts may arise. The existence of a critical mass of chief executives and the scarce and high-quality business information which they command is the principal external economy which is provided by world cities. Instead of the manufacture of goods, such places are now primary centres for the production of information.

Some insight into the nature of business information exchange in London was provided by Goddard in 1973. In this pioneering work, patterns of inter-firm linkage were monitored and analysed using data on telephone calls recorded in contact diaries. Six major clusters of activities were identified, concerned with commodity trading, publishing and business services, civil engineering, fuel and oil, official agencies, and banking and finance. Communication among these groups accounted for nearly two-thirds of the pattern of linkage within the area. Two particular features of the contact network which enhanced its value to decision-makers were noted. The first was that not all firms within a specialist group were linked together by reciprocated information flows, for in addition to within-cluster linkages there were also important connections between groups. The second was the hierarchical nature of many linkages which, combined with the interdependency of the various clusters, means that the individual contacts formed part of a highly integrated and structured system. The strength and complexity of connections suggested that the advantages of locating in the area were substantial and self-reinforcing. Organisations which open head offices in central London can clearly plug into a rich and diverse network of contacts which is likely to enhance the effectiveness of their information-gathering and decision-making processes. Conversely, firms which leave the area are likely to suffer 'communications damage' and to find that the effectiveness of their strategic orientation functions declines.

INCREASING GLOBAL REACH

The growth and concentration of global capital is made possible and sustained by complex flows of people and ideas. Information is the key business resource and it is by controlling the flow of instructions, ideas and data to regional offices, branch plants, affiliates and subsidiaries, that headquarters personnel are able to manage their global empires. Recent developments in telecommunications and transport facilitate and underpin this command and coordination function. They permit what were once separate and dispersed economic activities to become integrated and concentrated functions. Together they have eradicated traditional barriers to interaction, such that time and space have collapsed to a point. Such points are world cities.

Table 19 Advances in telecommunications

Advance	Year
First submarine telegraph cable, North Atlantic	1866
First submarine telephone cable, North Atlantic	1956
Formation of INTELSAT consortium	1964
First commercial communications satellite	1965
Formation of INTERSPUTNIK	1974
Formation of ARABSAT	1976
First fibre-optic cable, North Atlantic	1988
First fibre-optic cable, North Pacific	1988
First international Integrated Services Digital Network (ISDN) link	1989

The creation of a network of global business linkage is almost wholly attributable to advances in telecommunications (Table 19). Although the telegraph and the telephone were invented in the nineteenth century, it is only in the last forty years that telecommunications have become media of instantaneous and low-cost world-wide communication. The potential for global linkage was demonstrated by the laying of the first intercontinental telephone cable across the Atlantic in 1956 and the launch, in 1957, of Sputnik 1. It was facilitated in 1964 with the establishment of INTELSAT, an international consortium, led by the USA, which aimed to create a network of communications satellites. In 1965, the sixteen members of the cooperative launched Early Bird 1, the first commercial communications satellite. Today, INTELSAT has 119 members and links more than 170 countries, territories and dependencies world-wide through a network of thirteen operational satellites and over 700 ground stations (Akwule, 1992; p. 52). It is paralleled and complemented by the global communications system created since 1974 by the sixteen members of INTERSPUTNIK, the satellite consortium in the former Soviet Union. In the Middle East, the twenty-two members of the Arab League of Nations are further served by the two ARABSAT satellites.

Global communication has further been facilitated since 1988 by the introduction of fibre-optic cable systems which have vastly increased transmission capacity between principal business centres. Moreover they possess considerable amounts of redundancy in the form of multiple transmission paths and so, in the event of systems failure, downtimes are minimised (Warf, 1989). In common with earlier international telecommunications innovations, the first intercontinental fibre-optic cables were built to carry heavy traffic on the North Atlantic. The first cable (TAT-8) with a capacity of 8,000 voice equivalent circuits was completed in 1988 and a second in 1992 (Langdale, 1991). A North Pacific public fibre-optic cable system linking California with the Philippines, Hong Kong and Japan via Hawaii and Guam, was completed in 1988. While at present fibre-optic systems duplicate and supplement satellite services, the large carrying capacity and relatively low cost of cables means that in the future they are likely to benefit internal communications in the world's poor nations.

The most recent innovation is the development of an integrated services digital network which allows customers to send and receive high-speed, high-quality voice, data, image and text, in any combination, over public telephone lines. International ISDN links were first established in 1989 between the United Kingdom and France, USA and Japan, while in 1990 service began to Australia, the Far East, Scandinavia and Germany. ISDN customers benefit from a wide range of services and clearer, quieter and more reliable connections. The main advantage is that data can be transmitted faster, and without the need for a modem to convert signals from analogue to digital mode and back again. ISDN creates the basis for an integrated electronic office in which data, sound and vision can be transmitted simultaneously to major business centres across the globe.

The development and introduction of such technologies have been helped in recent years by the deregulation of telecommunications industries across the world (Moss, 1988). The most important was the abolition in 1984 of the American Telephone and Telegraph (AT&T) monopoly, which created new opportunities for competitors such as MCI and US Sprint (Warf, 1989). In Great Britain, the privatisation of British Telecommunications and the entry into the market of the Mercury Corporation was a similar process. In 1987 the Japanese government licensed three new companies to compete with Nippon Telegraph and Telephone. The deregulation of the telecommunications industry was important because telephony is the leading form of telecommunications at the international scale. The major consequence was a reduction in long-distance call rates, from which international voice and data transmission services benefited the most.

The emergence of world cities as financial centres has largely been made possible by telecommunications. London, New York and Tokyo have long been regional financial centres, but the introduction of international telecommunications and computer systems enabled them to develop global banking and financial trading functions. Of particular importance is the interlinkage of dealing rooms through dedicated and secure telecommunications networks so that currencies, stocks and shares and commodity prices and volumes in every significant business centre are simultaneously displayed on computer screens around the world. On this basis, twenty-four hour global trading can be conducted at the touch of a button, with financial settlements being made by electronic funds transfer. Computers are commonly programmed to buy and sell automatically if prices fluctuate beyond specified limits. London is especially well placed to dominate the global financial market as it lies between the Eastern Standard and Far Eastern time zones. Telecommunications enable a London-based dealer who starts work at 0600 hours to catch the end of trading on the Tokyo exchange, to trade all day in London, and to conduct business for several hours in New York.

It is important to emphasise that telecommunications create and enhance rather than erode world city functions. Although telecommunications enable interaction to take place without participants travelling to central meeting places, they are not appropriate media through which to conduct the types of 'orientation' meeting that take place in world cities. The purpose of such meetings is to evaluate options, to negotiate deals and to take decisions. They typically involve top-level personnel and their advisers. It is essential for participants to be present in person since the aim is to float ideas, to gauge reactions, to cajole, to persuade and to decide. None of these activities can adequately be performed remotely. The face-to-face activities that take place in boardrooms and on the dealing floors in major financial centres have not been and are unlikely to be made obsolete by new technology; rather, technology has extended the global reach of those who transact such business and so has reinforced the status of world cities.

The pre-eminence of world cities is maintained and enhanced by the way in which new communications technologies are introduced. World cities benefit most from advances in telecommunications because, as established locations for global business, they are the places which first receive and derive the advantages of new services and applications. Telecommunications, in common with many innovations, diffuse hierarchically through urban systems. They are initially made available in major cities where traffic levels, revenue and profits are greatest, and they subsequently spread down and

159

outwards to successively lower-order centres. Early adopters gain the greatest locational advantages. They benefit in two main ways: first, by becoming more easily accessible to each other and, second, by experiencing a reduction in their communication costs as a result of the lower tariffs that are commonly associated with new technologies. Comparative advantage is compounded by the high level and frequency of innovation that has characterised the telecommunications sector in recent years. The latest transmission and terminal technologies are introduced before their predecessors have diffused completely through the urban system. Today, the complex world of global telecommunications includes systems that have been in use for many decades alongside other innovations that are barely ten years old. World cities are the principal beneficiaries since they are typically several technologies ahead of competing lower-order centres.

This pattern of spatial diffusion and its consequences are demonstrated historically by the spread of international subscriber dialling (ISD) in the United Kingdom (Clark, 1979). In 1972 about one-third of UK telephone subscribers had access to ISD, but they were concentrated in the central areas of the six largest cities. Medium-sized centres were linked to the service in 1974 but the conversion of the whole country was not completed until 1984. The diffusion of ISD followed the same pattern as that of subscriber trunk dialling which began some fifteen years earlier (Clark, 1974). The implications for the location and conduct of business were important and far-reaching. For about a decade, London had direct dial access to the principal business centres in Europe and North America, whereas it required an operator to connect international calls from many centres in rural and provincial Britain.

An integrated world economy dominated by world cities is largely the creation of global telecommunications as there have been few comparable improvements in business travel. The last major technological advance was the introduction, as long ago as 1976, of supersonic Concorde aircraft on the North Atlantic routes. This reduced the flying time between London Heathrow and New York's John F. Kennedy airports from eight hours by subsonic services, to four hours. Travel times on very long haul routes, including those between London, New York and the principal Asian business centres have recently been reduced because of the introduction of non-stop 747–400s, but otherwise the airline industry has been on a performance plateau for the last twenty-five years. Travel times between city centres have probably risen because of road traffic congestion and the time now required for check-in, security clearance and baggage reclaim. Regular long-haul business travel is tolerated by most but relished by few, to the relative advantage of telecommunications links. Further significant reductions in air travel times are not envisaged until the introduction of hypersonic aircraft in the early twenty-first century (Janelle, 1991: p. 31).

The central importance of telecommunications in supporting world city functions is reflected in the changing design and construction of office buildings. Global financial companies need vast open spaces to accommodate their dealing rooms and trading floors, together with high ceilings to carry the cabling for telephones and computers and the ducting for air conditioning. Such requirements are met by the construction of dense deep-plan buildings which cover the whole of the site. They typically feature central courtyards and atria which bring daylight into their centres. Such 'ground-scrapers' contrast markedly with the 1960s thin-clad high-rise office blocks with their inadequate servicing, forests of internal columns and low ceilings (Williams, 1992). Their proliferation in the urban landscape provides powerful visual evidence of the fundamental importance of telecommunications to the command and control functions of world cities.

TOKYO AS AN EXAMPLE OF A WORLD CITY

The roles played by the institutions of global capitalism, their geographical concentration, and transport and telecommunications are well illustrated by the rise of Tokyo as a world city. In contrast to London and New York, Tokyo was a late entrant into the world-city league. Up to the 1930s Japan was a bipolar country, with Tokyo and Osaka competing for national pre-eminence. Significant investments were made at the time in colonies and other territories in East Asia and it was imperialist aspirations that were responsible for Japan entering the Second World War. Defeat and subsequent occupation gave rise to a geographical restructuring in the form of a concentration of economic activities around Tokyo, that laid the foundations for national and subsequently world city status. During and immediately after the war many Osaka-based industrial corporations relocated their headquarters to Tokyo as the centre of political power, as did many non-Tokyo banks. A similar trend occurred in higher education and research, such that the University of Tokyo emerged as the dominant educational institution (Markusen and

Gwiasda, 1994). The fact that research-led microelectronics is the leading sector in the national economy meant that Tokyo became the principal manufacturing centre in Japan.

Its continued success as an industrial centre owes much to the development and exploitation of systems of flexible production which enable manufacturers to switch rapidly from one product to another in response to changes in domestic or world markets. Such systems extend beyond the availability of flexible machinery and skilled labour to include networks which bind research and development specialists, manufacturers, distributors, government agencies and service suppliers in a highly responsive production arrangement (Fujita, 1991). Flexible production is facilitated by and in turn supports a distinctive industrial structure in which most manufacturing is undertaken in small factories. Most large plants with over 300 workers have moved out of Tokyo in recent years and the number of small plants has increased proportionately. 'If the total number of plants in 1960 is indexed at 100, the number of large plants fell to 53.6 by 1983, but the number of small plants with less than 299 workers rose sharply to 185.8' (Fujita, 1991: p. 275). Presently, some 84 per cent of plants employ between four and nineteen workers. The concentration of small flexible producers gives Tokyo a special character as a manufacturing city. It is a centre of innovation, a place where new industries and hybrid industries originate.

Tokyo became a world city as a consequence of the transnationalisation of Japanese commercial capital which was generated through manufacturing. In the early post-war period Japan's foreign investments were aimed primarily at securing raw materials such as oil, wood and pulp for domestic industrial production. After about 1970, capital exports were liberalised and gave rise to two distinct waves of capital investment. The first was in factories in nearby Asian and Pacific countries with low labour costs, to manufacture electronic components and semiconductors. The second was in advanced Western economies, including the United States, in the form of plants through which Japanese electrical goods and cars could be imported and assembled.

As globalisation proceeded, Tokyo became the principal centre in Japan for international trade and finance. A rapid increase in the number of corporate headquarters in the city occurred in the 1980s as corporations sought to locate in the central business district for symbolic reasons, to gain international competitiveness, to have access to information sources, and to deal with international trading and transactions. The wealth which they generated was recycled overseas through foreign securities and currency markets by Japanese banks, foreign financial institutions which were attracted to Tokyo, and by branches of Japanese banks in other world cities, principally London and New York. The presence of Japanese corporations in the global economy attracted foreign direct investment and employment by American and European companies, most of which was concentrated in Tokyo. Some 66 per cent of the 70,000 foreigners in Japan in 1988 lived in the city. Tokyo's status as a world city derives from its concentration of corporate head offices, financial institutions and supporting producer service organisations which exceeds that of any centre in Asia.

Although the translation of domestic economic power into global influence is a process that occurred in most world cities, Tokyo is different to its principal rivals in several important respects (Machimura, 1992). First, it is mainly a world economic centre which is unsupported by a political or military hegemony of the state. Second, it is heavily dependent upon global communications systems because it is physically far away from the Western countries that are the traditional centres of the world economy. Advances in telecommunications and transport, assisted by the success of the Japanese economy in electronics, were indispensable for it to overcome its geographical disadvantage. Third, Tokyo was for many years closed to the beneficial influx of foreign immigrants. For Machimura (1992), New York and London are world cities because of the political, military and cultural hegemony of the state. Tokyo is a world city because of its global network of economic activities.

CONCLUSION

The cities which are identified and analysed in this chapter occupy the top tier in the global urban hierarchy. They are distinguished not by their size or status as national capitals, but by their specialisms and the roles which they perform. Such is the degree of concentration of the headquarters of transnational corporations, the head offices of banks, finance dealers and the supporting producer service organisations that they function as command and control points for global capitalism. Their true status and the way in which this is conveyed in the form of a descriptive label are matters of legitimate and unresolved debate. What is clear, however, is that the individuals and institutions which are located within them exert a disproportionate and, according to some analysts, decisive influence on the shape and structure of the contemporary urban world.

Many places lay claim to world city status but only New York, Tokyo and London have the critical mass of corporate, financial and service functions to satisfy the criteria for inclusion in this elite group. Other, lesser centres have some global functions but these cities have the largest number and the most powerful. Heads of corporations and governments, and their key staff, cluster together in world cities so as to generate and to benefit collectively from information and to further the interests and activities of their businesses by being at the centre of world markets. They manipulate and play these markets via telecommunications and computer services which give low cost and instantaneous global reach. Dominant cities are associated with the world-systems of the past, as Chase-Dunn (1985) has emphasised, but the world cities today are the first to perform a global role. Their emergence reflects and underlines the extent to which capitalism, and hence global urban development, is shaped by a minority of decision-makers working out of a small number of key locations.

World cities dominate the settlement hierarchy in a highly developed urban world. They are the control points for a capitalist system which has helped to concentrate large and growing numbers of people in urban places. Many live in rapidly expanding mega-cities, especially in the developing world. Although the present pattern appears to be viable, although unjustly structured, there are many concerns as to its future. How it is likely to evolve and whether it is sustainable are topics which are addressed in the final chapter.

RECOMMENDED READING

Brunn, S. D. and Leinbach, T. R. (1991) *Collapsing Space and Time*, New York: Harper Collins.

An edited collection of papers on all aspects of the impact of telecommunications upon urban economy and society, including their contribution to global interdependence and the rise of world cities.

Daniels, P. W. (1993) *Service Industries in the World Economy*, Oxford: Blackwell.

A useful overview of the organisation and growth of financial and producer services, and their locational preferences.

Hall, P. (1984) *The World Cities*, London: Weidenfeld & Nicholson.

The third edition of a work published originally in 1966, in which the concept of world cities is introduced and evaluated, and a number of detailed case studies are presented.

King, A. D. (1990) *Global Cities: Post-imperialism and the International-isation of London,* London: Routledge.

This book combines a scholarly analysis of the development of the world economy and the role of London as a colonial, imperial and world city. The treatment of the social and physical consequences of London's world city status is comprehensive and highly detailed.

Knox, P. L. and Taylor, P. J. (eds) (1994) *World Cities in a World System,* Cambridge: Cambridge University Press.

This book comprises a set of research papers, originally presented at an international conference, on the world city hypothesis and the world system. All aspects of the current debate on world cities and their role in the global economy are discussed.

Sassen, S. (1991) *The Global City: New York, London, Tokyo,* Princeton NJ: Princeton University Press.

An analysis of the growth and characteristics of the three principal world cities.

Sassen, S. (1994) *Cities in a World Economy,* London: Pine Forge.

A sociological examination of the urban impact of economic globalisation and the rise and role of world cities.

8

THE FUTURE
URBAN WORLD

It is appropriate in a book which adopts a broadly historical approach to conclude with a brief consideration of how the urban world may evolve in the near future. Forecasting population levels, distributions and socio-economic conditions in a single country is a difficult enough task, and attempts to undertake this sort of exercise at the global scale can only yield predictions which are little more than guesstimates. This chapter is therefore grounded in speculation rather than in detailed analysis. What is clear, however, because they are products of long-term and deep-seated processes which have yet to run their course, is that urban growth and urbanisation will lead to further significant urban development. Most will be in those parts of the world which are presently classified as developing. Even over the next quarter of a century the changes which are expected are staggering. Extrapolation of current trends suggests that the number of people who presently (1996) live in urban places is likely to double by the year 2025.

The scale of urban development which these figures imply raises important questions as to whether such a geographical pattern can be supported. It is difficult to imagine a world with twice as many urban residents as today, and it is important to focus on issues of maintenance and sustainability. Cities are elements in global economic and environmental systems which are both vulnerable and fragile. Although they represent a highly efficient use of space and provide unrivalled opportunities for production and social inter-action, they consume prodigious amounts of finite resources, far more than a rural population of equivalent size. There are grave doubts as to whether future cities can be sustained in economic terms, how the population can be fed and how they can generate and distribute sufficient wealth to support their residents at acceptable standards of living. A parallel concern is with the ecological implications of further urban development. Cities are widely seen as being parasitic, in that they draw air and water from the natural environment and generate large quantities of pollution and waste. Little is reused and recycled. Many would argue that their emissions are progressively destroying the global environmental systems upon which life on the planet, and hence their own viability, depends. The prospect for further massive urban development necessarily focuses attention upon the implications for the environment and whether the urban future is sustainable.

Such questions raise a further set of issues concerning the need for regulation today so as to ensure the continuation of cities into the future. If urban life is to be sustained much beyond the present century, then steps must be taken now so as to prevent further damage to the environment and to bequeath adequate resources to succeeding generations. Action is required at the global scale to cut harmful emissions and to prevent indiscriminate and unnecessary exploitation of scarce resources. Within national boundaries, there is a need to deal with local sustainability issues including urban servicing and waste disposal. Agendas for such intervention are presently emerging although these are more high-level statements of intent than examples of concerted and effective action. The urban future is likely to depend as much upon the success of international agencies and governments in shaping urban development as it is on the unregulated growth and redistribution of the population.

THE URBAN FUTURE

The direction and scale of contemporary urban growth and urbanisation point to the emergence by 2025 of an urban world that will bear little resemblance to the urban present. This much is certain, but filling in the detail by country and by region is problematical because of the deficiencies of the data, which have been emphasised throughout this book and which are addressed in the Appendix. A related difficulty is methodological and concerns the interpretation of past trends and the ways in which they are used in forecasting. Urban development is too sensitive to economic, social and environmental change to predict more than one or two decades into the future (Hardoy and Satterthwaite, 1990). An example of the very different pictures which can emerge is provided by the projections for city populations in the year 2000 that were published by the

Table 20 Examples of changing projections for city populations, in
millions, by the year 2000

| City | UN projection for population in the year 2000 | | | |
	1973–5	1978	1982	1984–5
Mexico City	31.6	31.0	27.6	25.8
São Paulo	26.0	25.8	21.5	24.0
Calcutta	19.7	16.4	15.9	16.5
Rio de Janerio	19.4	19.0	14.2	13.3
Shanghai	19.2	23.7	25.9	14.3
Bombay	19.1	16.8	16.3	16.0
Seoul	18.7	13.7	13.5	13.8

Source: United Nations Population Division, Department of International Economic
and Social Affairs

United Nations between 1973 and 1984 (Table 20). The projected
population of Mexico City in the year 2000 was 31.6 million in the
Population Division's 1973–5 assessment, but was much lower at
25.8 million in the 1984–5 forecast. Similarly, the 1973–5 projec-
tion for the population of Beijing in 2000 was 19.1 million, but
this had been revised downwards to 11.3 million in 1984–5. The
scale of revision by the leading statistical agency underlines the innate
difficulties which are involved in long-term urban forecasting.

The most recent estimate by the United Nations (1991) is that
by the year 2025 there will be some 5.5 billion people, out of a
world population of 8.5 billion, living in urban places. This future
urban population is roughly the same as the total population of the
world today. Some 4.4 billion will be living in towns and cities in
what are presently classified as developing countries. The population
of China's urban places will be close to 1 billion and India's is
expected to be some 740 million (Figure 23).

Urban growth will be accompanied by increased urbanisation.
Some 65 per cent of the world's population is expected to be urban
by the year 2025. It follows from the analysis and discussion in
Chapter 4 that this increase will occur principally because of the
urbanisation of the population across large parts of Africa and Asia.
These regions will be most radically affected by urban development,
both urban growth and urbanisation, in the next quarter century.

The most striking feature of the predicted urban geography of
the year 2025 is the uniformly high level of urban development in
the Americas (Figure 24). The population of all of the principal
countries of North, Central and South America is expected to be

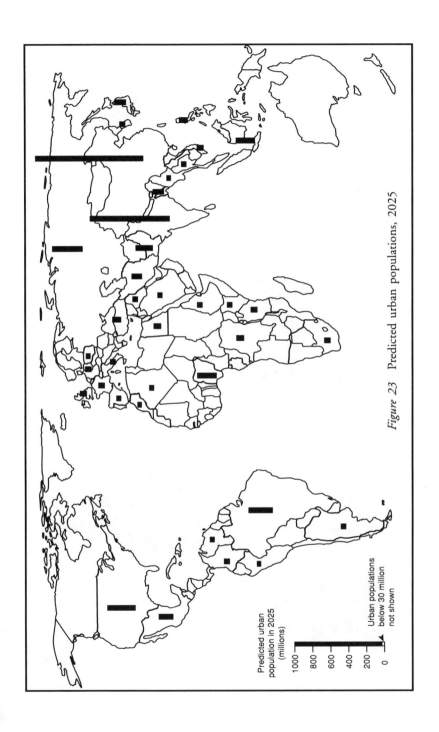

Predicted urban population in 2025
(millions)

1000
800
600
400
200
0

Urban populations
below 30 million
not shown

Figure 23 Predicted urban populations, 2025

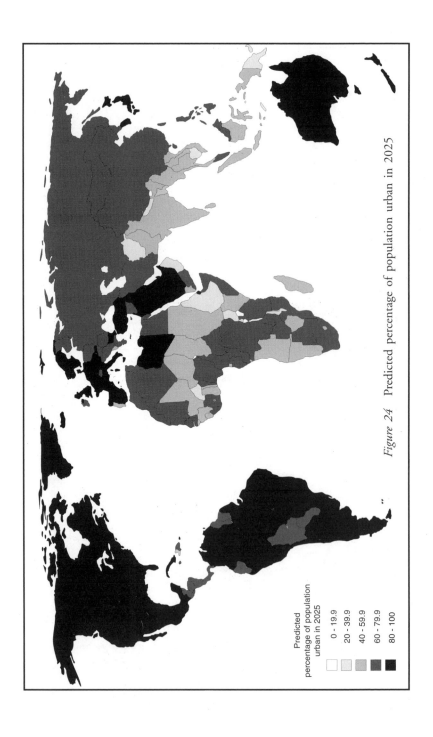

Figure 24 Predicted percentage of population urban in 2025

Predicted
percentage of population
urban in 2025

0 - 19.9

20 - 39.9

40 - 59.9

60 - 79.9

80 - 100

over 60 per cent urban and in most it will be in excess of 80 per cent. The Americas are presently highly urbanised so this change represents a consolidation of existing patterns. Similarly high levels of urban development are anticipated in Australia, Japan, parts of the Middle East, North Africa and most of Europe. Urbanisation levels in excess of 60 per cent are expected across the whole of Asia north of the Himalayas.

The United Nations forecast suggests that levels of urban development in Africa and southern Asia will be very much higher than today, but will vary considerably from country to country. Although the population in most countries will be more urban than rural, the proportion living in towns and cities in Burundi, Malawi, Rwanda, Ethiopia, Uganda, Burkina Faso, Afghanistan, Nepal, Bhutan, Cambodia, East Timor and Papua New Guinea is expected to be less than 40 per cent. Such countries have yet to go through the phase of rapid urbanisation that is a characteristic feature of the cycle of urban development (Figure 12). They will be the world's last remaining rural territories. At 51 per cent, India is expected to be only marginally more urban than rural in 2025.

If migration continues as it has in the past, and there is presently no suggestion that it will not, many who move from rural areas to urban places will go to the largest cities. The number of mega-cities is expected to rise to 28 in 2000 and to be more than 60 in 2025 (United Nations, 1991). Strong patterns of primacy already exist in many African countries and these are likely to be reinforced rather than reduced by urbanisation and urban growth. An important consequence of continued primacy is the further polarisation of the settlement hierarchy. The problem will be not so much the over-concentration of the urban population in one city, as the wide dispersal of people in settlements too small to support even basic services and non-agricultural economic activities. Even by 2025, in the least urbanised countries in Africa there are expected to be too few urban centres of any size. Many urban places will have population thresholds that are too low to support the variety of services needed to stimulate the commercialisation of agriculture, to meet the basic needs of the rural poor, and to increase the productivity and income of the rural population. A widening of the gap between town and country is expected to be one of the major consequences of the urban transition in Africa over the next thirty years.

The urban geography of the developed world is likely to be very different. Here, urban populations are presently high and the principal shifts will take place among cities. Rather than a concentration in a small number of large cities, which is the current pattern, the population is expected to be more evenly spread across many smaller centres. Cities of about 200,000 are likely to be the most attractive, as they are large enough to sustain an acceptable range of services without the congestion and pollution that are associated with life in the mega-city. As the benefits of centrality and agglomeration lessen further, many people will be drawn to places which may be too small, in future terms, to merit designation as urban. The spatial structure of large cities is likely to be transformed by shifts in population and economic activity. Central area densities are likely to fall significantly as people and businesses move to the suburbs and beyond. Many of the areas which are vacated will become parkland and open space, so that cities will take on a 'doughnut' form. In the longer term, however, it is possible, as Table 3 suggests, that central area populations will rise as cities begin to reurbanise. Decentralisation of population at the local scale, and deconcentration at the national level will significantly reduce urban/rural differences, so producing a 'rurban' arrangement. The expected trends in different parts of the world will lead to a progressive inversion of the contemporary urban pattern at the global scale. A small-city and rural orientation will increasingly characterise the landscape in developed economies but strongly urban-centred mega-city societies will emerge and predominate in the developing world.

ISSUES OF SUSTAINABILITY

The preceding forecasts paint a disturbing picture of major population increase and further massive urban development in the first half of the next century. They point to a future in which the population in most parts of the world will live in urban places and in many cases in mega-cities. Such is the scale of urban development which is predicted, however, that it raises questions as to whether urban development of this magnitude can be sustained. It is difficult, given the current pressures on resources and the environment, to see how urban populations can double over the next twenty-five years without some form of economic or ecological breakdown. How will the urban population be fed, and what effects will mass concentration of population have upon the global environment and upon local ecosystems? Posing such questions is easy but arriving at plausible answers is much more difficult. The debate on sustainability is comparatively new and is strongly infused with speculation and conjecture. Hard information is lacking and the significance of such scientific evidence as is available, as for example on the scale of ozone depletion and global warming, is much disputed. Particular difficulties surround the availability of data for developing countries where environmental monitoring is in its infancy. Such problems do not of themselves negate discussion but they mean that it must be conducted at a very general level. It follows that few clear conclusions can be reached that might assist urban planners and managers.

A general concern with sustainable development emerged as a key issue with the heightened environmental awareness of the late 1980s. It gave rise to the emergence of 'green' parties in many developed countries in which the membership was committed to forcing environmental issues to the centre of the political debate. Evidence for global warming, the depletion of the ozone layer, and the detrimental effect of acid rain highlighted the need for urgent action to prevent further environmental degradation. Sustainable development can be interpreted in a number of different ways; indeed Pearce, Markandya and Barbier (1989) in an appendix to their book *Blueprint for a Green Economy* quote twenty-four separate definitions. The most widely accepted is that of the report of the World Commission on Environment and Development (WCED, 1987), also known as the Brundtland Commission, which defined sustainable development as 'development which meets the needs of the present without compromising the ability of future generations to meet their own needs' (p. 43). It further elevated the notion of sustainable development to the level of an operational concept which embodied the principles and values which it saw as desirable and necessary so as to deal effectively with the crisis of the environment and the development process. Emphasis was placed on the need for action today so as to provide for economic and ecological viability tomorrow.

Haughton and Hunter (1994) argue that the concept of sustainability involves three major principles. The first is that of 'intergenerational equity' and concerns the legacy which is left to future generations. It argues that the success of cities in the future depends to a large extent upon the assets and resources that are available and it is therefore incumbent upon the current generation not to indulge in indiscriminate and wasteful consumption. A sustainable future requires that national capital assets of at least equal value to those of the present are passed on to succeeding generations.

The second is that a fair and equitable use of present resources is clearly necessary and this is enshrined in the principle of 'social justice'. Some form of central control over access to and use of resources is implied. The fact that both resources and consumption are widely distributed and are interdependent means that such management must be at a broad scale. A third precondition for sustainable development is that of 'transfrontier responsibility' insofar as key issues such as pollution, waste disposal and climatic warming are not constrained by national or regional boundaries but are essentially global in cause and consequence.

Against this background, Blowers (1993) identifies five fundamental goals that should guide all decisions concerning future development so as to ensure sustainability. The first concerns conservation and involves the need to ensure the supply of natural resources for present and future generations through the efficient use of land, less wasteful use of non-renewable resources, their replacement by renewable resources wherever possible, and the maintenance of biological diversity. The second concerns the use of physical resources and their impact on the land. It seeks to ensure that development and the use of the built environment is in harmony with the natural environment and that the relationship between the two is one of balance and mutual enhancement. A third goal is to prevent or reduce the processes that downgrade or pollute the environment and to promote the regenerative capacity of ecosystems. The final two goals are social and political in character. The aim of goal four is to prevent any development that increases the gap between rich and poor and to encourage development that reduces social inequality. The final goal is to change attitudes, values and behaviour by encouraging increased participation in political decision-making and in initiating environmental improvements at all levels from the local community upwards.

A specific focus upon the sustainability of cities arises because it is at the urban level that many environmental problems are sourced and where they are experienced with greatest intensity (Organisation for Economic Cooperation and Development, 1990). Cities are both great consumers and degraders of the natural environment. They extract far more from the environment than they return. The concern for sustainability arises out of a recognition that urban development is a linear process (Girardet, 1990): 'food, fuels, construction materials, forest products and processed goods are imported into the city from somewhere, never mind where, and when they are finished with they are discarded, never mind how' (p. 7). For example it is estimated that only 15 to 20 per cent of urban-sourced nitrous oxide pollution falls inside the city, the rest contaminating rural areas often many kilometers away (Alcamo and Lubkert, 1990). Girardet points out that this linear system is profoundly different to nature's own circular metabolism, in which every output is also an input which renews and sustains life. As the urbanisation level increases, this imbalance has massive implications for the well-being of the world's forests, soils, water courses and atmosphere. Such degradation in turn threatens the cities which contribute to it.

Some indication of the highly detrimental effect of cities is provided by measures of the levels of suspended particulate matter such as soil, soot, smoke, metals and acids which are found over cities as opposed to adjacent rural areas. Goudie (1989) reports the average concentration of suspended particulate matter found in the commercial areas of a number of cities as 400 micrograms per cubic metre in Calcutta, 170 in Madrid and Prague, 147 in Zagreb, 43 in Tokyo, and 24 in London and Brussels. These values compare with concentrations of less than 10 micrograms per cubic metre in rural areas.

An estimated 1.4 billion urban residents world-wide are exposed to averages for suspended particulate matter or sulphur dioxide (or both) that are higher than the levels recommended by the World Health Organisation. Research reported by Hardoy, Mitlin and Satterthwaite (1992: p. 79) underlines the severity of the pollution which hangs over many Third World cities. For example, in Shanghai there are seven power stations, eight steel works, 8,000 industrial boilers, 1,000 kilns, 15,000 restaurant stoves and one million cooking stoves, most using coal with a high sulphur content. In 1991 the annual average concentration of sulphur dioxide in the urban core was more than twice the recommended level. The annual average for total suspended solids was more than four times that recommended. Similar situations exist in São Paulo and Bangkok, where suspended particulate matter routinely exceeds recommended levels at all the monitoring stations in the city. The effects of emission may be compounded locally by physical geography. Santiago has one of the highest levels of air pollution, because the surrounding mountains impede natural ventilation.

Although the general pattern of exchanges is the same in all cities, the magnitudes involved vary widely. Cities do not contribute to environmental damage on an equal basis. Those in developed countries consume disproportionate amounts of resources and in return generate excessive amounts of waste. It is estimated that some 25 per cent of the world's population lives in the highly urbanised countries of the developed world but they account for 70 per cent of the world's energy consumption, 75 per cent of metals and 85 per cent of wood (Tolba and El-Kholy, 1992). The major cities of the developed world draw upon the ecological capital of all other nations to provide food for their economies, and land, air and water in which to discharge their waste products. The Sears Tower in Chicago is a powerful visual symbol of the differential demands of cities: 'this ugly monster uses more energy in twenty-four hours than an average American city of 150,000, or an Indian city of more than one million inhabitants' (Hahn and Simonis, 1990: p. 12). Research reported by Haughton and Hunter (1994) suggests that urban residents in developed countries generate an average of 0.7–1.8 kg of domestic waste daily, compared to 0.4–0.9 kg daily in developing countries. Measured in per capita terms, urban residents in the USA and Australia are estimated to generate carbon dioxide emissions which are up to 25 times the levels in Dhaka, Bangladesh (Hardoy, Mitlin and Satterthwaite, 1992).

The concern for the sustainability of cities has been expressed at two levels. The first is global and involves a wide range of issues surrounding the long-term stability of the earth's environment and the implication for cities. It is clear that the world's cities cannot remain prosperous if the aggregate impact of their economies' production and their inhabitants' consumption draws on global resources at unsustainable rates and deposits wastes in global sinks at levels which lead to detrimental climatic change. The second is local and involves the possibility that urban life could be undermined from within because of congestion, pollution and waste generation and their accompanying social and economic consequences. These different concerns focus attention upon the need for intervention at an international scale by governments working together on agreed programmes, and at the domestic level by city authorities addressing the local sustainability issues over which they can exercise some control.

There is growing evidence from climatological research that the earth's atmosphere is being degraded to an unacceptable extent, with serious implications for life on the planet. There are particular concerns for the well-being of the global climate which it is believed is being threatened by the depletion of the ozone layer and by atmospheric warming. The layer of ozone which exists in the upper atmosphere is of vital importance in the global energy balance because it reduces the amount of harmful solar radiation which is received at the earth's surface. Ozone occurs when oxygen reacts with ultra-violet light to give a molecule of three oxygen atoms, and its concentration and distribution within the lower stratosphere remain roughly constant under normal conditions. There is growing evidence, however, that the natural cycle of creation and breakdown of ozone in the upper atmosphere has been seriously interrupted by certain compounds, especially chlorofluorocarbons (CFCs). These chemicals are widely used in refrigeration, aerosols, packaging and cleaning, and levels of production have increased significantly in recent years as these applications have grown. CFCs live for a long time and have accumulated in large concentrations in the lower stratosphere where they are thought to have caused a general thin-ning of the ozone layer and the appearance of ozone holes. The most extensive hole is that over the Antarctic and there are indications that it is increasing in size and may now cover parts of Australasia and South America. Severe ozone depletion has also been observed during winter months in middle and high latitudes in the northern hemisphere (Tolba and El-Kholy, 1992).

It is widely believed that ozone depletion has led to a rise in ultra-violet radiation, which in turn is affecting human health and is threatening many natural and semi-natural ecosystems. The magni-tudes are difficult to establish but a conservative estimate is that a 1 per cent reduction in stratospheric ozone leads to a 3–4 per cent increase in non-melanoma skin cancers (Turner, Pearce and Bateman, 1994). Sunbathers in areas where the ozone layer is thinnest are at greatest risk. Ozone depletion is also thought to cause eye damage and to suppress people's immune systems. The yield of some commer-cial food crops may be reduced, as may fish stocks (Tolba and El-Kholy, 1992).

A second set of pollution-related changes is believed to be raising average temperatures across the world, with far-reaching implications for climate, global sea levels and the functioning of local ecosystems. Global temperatures are regulated by a layer of natural 'greenhouse gases' in the atmosphere, including water vapour, carbon dioxide, methane and nitrous oxide. These trap longwave radiation emitted by the earth and reflect some of it back to the surface in the form of heat. There is now compelling evidence that a build-up of pollution in the atmosphere has compounded the greenhouse effect and has caused long-term global warming (O'Riordan, 1989). The principal pollutant is carbon dioxide, which is produced during the burning of fossil fuels, but CFCs are important absorbers of long-wave radiation as well. Atmospheric levels of carbon dioxide have risen by around one quarter over the past two centuries, with about half of the increase occurring in the last forty years (Kelly and Karas, 1990). Historically, emissions were higher in the developed world but the fastest growth today is occurring in developing countries in association with coal-fired heating, inefficient power stations, and the rise in the number of motor vehicles.

Although the scientific evidence on the scale of change is equiv-ocal, global warming is widely seen as an increasing long-term threat. The environment is finely balanced and a rise in average global temperatures of as little as 1 per cent could melt polar icecaps suffi-ciently to raise sea levels across the world by half a metre. Such a change would threaten densely populated areas on deltas and coastal plains. Many major cities, including New Orleans, Amsterdam, Shanghai, Dhaka and Cairo, are wholly or partly below present sea level and any rise would significantly increase the risk of flooding. Increases in temperature may also contribute to desertification and reduce agricultural production, especially in areas which are presently semi-arid (McMichael, 1993). Global warming is also predicted to lead to greater climatic fluctuations, accentuating summer tempera-tures and depressing those in winter. Such large-scale shifts could have particular consequences for cities in semi-arid environments where climatic conditions are presently marginal.

The threats posed to cities by a degradation of the global envi-ronment are potentially serious, but they are likely to accrue only in the long term. A more immediate possibility is that cities could be seriously undermined from within because of the sheer pressure of numbers on infrastructure and basic services. There are many concerns involved, including fears over the ways in which the built environment is evolving and the implications for the effective func-tioning of cities as economic and social systems (Haughton and Hunter, 1994). Many cities, especially in the developed world, are suburbanising to such an extent that their coherence and integra-tion is being compromised. Others, particularly in the developing world, are severely stretched to provide public services today, and large numbers of their residents lack basic utilities and amenities. The principal implications for sustainability can be illustrated by an examination of the very different problems of urban sprawl, water supply and waste disposal.

Land use is a focus of growing concern because of fears that cities are in danger of consuming too much of this locally finite resource and are evolving as spatial forms which are not sustainable. The area of cities increases as populations rise and this, combined with greater locational flexibility for individuals and industries, has led to urban sprawl in place of the high-density compact urban forms that are associated with the pre-automobile age. In some cities the space demands of the car account for almost one-third of the urban land area, rising to two-thirds in inner Los Angeles (McMichael, 1993). Areal expansion is widely seen as inefficient, as it is associated with high energy consumption and increased pollution. Evidence is provided by McGlynn, Newman and Kenworthy (1991), who ranked a number of cities from across the world into five groups, from large sprawling US-type cities with high automobile dependence, to compact cities where there is little reliance upon the car. A high positive correlation was found between urban type and environmental impact. Compact cities had fewest adverse consequences for the environment, but sprawling cities had major detrimental effects. Sprawl is further condemned because of its adverse effects on the countryside and because some observers believe that it leads to cities which lack social cohesion, dynamism and vibrancy (Unwin and Searle, 1991). Others point out that in the long term, urban sprawl is counterproductive. Many of the benefits of the car are short-lived, as rising levels of ownership and use soon lead to congestion and paralysis, undermining the urban structures which the car helped to create.

Although low-density living has many supporters, not least among those who enjoy the environmental attractions of suburbia, there is a widespread view that the physical expansion of cities needs to be checked. There is a limit, probably already exceeded in some countries, to the extent to which the built environment can be allowed to encroach on green land. Such restraint does not, however, necessarily imply a return to compact cities, as high-density living has many disadvantages including congestion, noise, and lack of open space. An alternative way forward is to build new ecological communities based on notions such as permaculture where the population would be self-reliant and self-sufficient (Orrskog and Snickars, 1992). Such proposals for environmentally efficient settlements are attractive to many, though it is uncertain how they could generate the wealth to support large numbers of people at standards of living which would be acceptable in the twenty-first century. Under present circumstances they seem far-fetched but they may gain currency if the functioning of cities is seriously undermined by urban sprawl.

Although there are many cities that have large areas, the principal sustainability issues in the developing world are those of service provision rather than urban sprawl. Urban growth has placed undue stress upon municipal authorities and many cities are deficient in even the most basic public services. The lack of a piped water supply is especially significant, as many health problems are linked to water. Millions of residents of Third World cities have no alternative but to use contaminated water, or at least supplies whose price is high and whose quality is not guaranteed. The problems are serious in almost all developing countries, as studies overviewed by Hardoy, Mitlin and Satterthwaite (1992) emphasise. They paint a depressing picture of levels of provision which, in the opinion of the authors, is more accurate than that portrayed in official government statistics. For example, although most of the population in Accra has access to piped water, the system is often not operational (Table 21). In Bangkok, Dar es Salaam and Kinshasa, over half of households are connected and the occupants of the remainder must either use standpipes or buy water from vendors. The situation in Dakar is worse, as only 28 per cent of households have private water connections, while 68 per cent rely on public standpipes and 4 per cent buy from water carriers.

A similar pattern of low but variable provision characterises sewage disposal and garbage collection. Hardoy, Mitlin and Satterthwaite (1992) estimate that around two-thirds of the urban population in the Third World have no hygienic means of disposing of sewage and an even greater number lack an adequate means of disposing of waste water. Most cities in Africa and many in Asia have no sewers at all, and human waste and waste water end up untreated in canals, rivers and ditches. Where sewage systems exist, they rarely serve more than the population that lives in the richer residential areas. Some 70 per cent of the population of Mexico City live in housing served by sewers, but this leaves some three million people who do not. In Buenos Aires it is estimated that the habitations of 6 million of the 11.3 million inhabitants are not connected to the sewer system.

The implications for health of these very low levels of public service provision need no elaboration. Contaminated water and the accumulation of waste are widely recognised as being major causes of morbidity and early mortality, especially among children, the elderly and low-income groups (Cairncross, 1991). They are,

Table 21 Estimates of levels of servicing in selected Third World cities

	Percentage of households estimated to be served by:		
	piped water	*central sewerage system*	*garbage collection*
Accra (Ghana)	100	30	10
Bangkok (Thailand)	66	2	80
Dakar (Senegal)	28	N/A	0
Dar es Salaam (Tanzania)	53	13	33
Jakarta (Indonesia)	33	0	60
Kampala (Uganda)	18	19	10
Karachi (Pakistan)	38	N/A	33
Khartoum (Sudan)	N/A	5	N/A
Manila (Philippines)	N/A	15	50
São Paulo (Brazil)	N/A	N/A	33
Kinshasa (Zaire)	50	0	N/A

Source: abstracted from Hardoy, Mitlin and Satterthwaite (1992)

N/A = not available

however, merely two of the many areas in which cities in developing countries are deficient and which threaten their viability. The fact that levels of public service provision are low today raises far-reaching questions as to how they can cope with the vast increase in numbers which is expected in the future.

Despite the largely pessimistic tone of the sustainability literature there are, however, some grounds for believing that the arguments have been overstated. The debate on sustainability is in its infancy and many of the points that are made lack detailed empirical evidence and need to be evaluated in the light of experience. Sceptics are keen to point out that the present concerns are merely the most recent in a string of doomsday predictions for the city, none of which have materialised. Arguments that the population grows more rapidly than food production, so leading eventually to widespread famine and social breakdown, have been expounded by a succession of writers from Malthus in the late eighteenth century to Meadows *et al.* (1972) in their highly influential work on the *Limits to Growth*. Although differences in growth rates exist, the evidence is that such analysts have consistently underestimated the capacity of farmers to raise outputs. There are famines in several parts of the world today, but these are distributional problems as there is no shortage of food overall. The Malthusian paradox is that the agricultural sector in many developed countries is contracting, under strong government pressure, in order to deal with overproduction. A crisis of food supply is likely at some stage in the future, but it seems some way off.

A different basis for some optimism about the future is the success of past attempts to deal with urban environmental problems. Most cities in the developed world are cleaner today than they were twenty years ago, and their residents enjoy higher levels of health and amenity as a result. Air quality is one area in which improvements have been dramatic. The reasons are both technological, involving the switch from coal to oil and gas as energy sources, and political, as governments have introduced and enforced clean air legislation. The drive for clean air continues as the burning of oil and gas create different types of pollutant, but these can now more easily be tackled at source by improving the efficiency of combustion at power stations. The example of improvements in air quality suggests, however, that there is some justification for believing that with appropriate intervention and direction, a sustainable future for cities is a realistic possibility.

MANAGING THE URBAN FUTURE

The overwhelming message in the sustainability literature is that the urban populations which are envisaged in the first section of this chapter can only be achieved and maintained through careful planning and management of resources at both global and local scales. Agreement is required among states to work together with other countries on a common agenda to rehabilitate and to protect the global environment. Action is also necessary to regulate urban development within national boundaries. A start has been made, in that international protocols on environment and development have recently been signed in which most nation states have entered into commitments to promote sustainable human settlement. The expectation is that governments will set environmental goals for cities and will take appropriate steps to reduce resource use and pollution. Some countries and interest groups within countries have already embarked on this course of action.

Attempts to provide for sustainable development by tackling environmental issues at the global level have been led by the United Nations. Examples of past initiatives include the UNESCO Man and the Biosphere programme, which was launched in 1971, and the Centre for Urban Settlements' HABITAT programme for sustainable cities (HABITAT, 1987). Agreements to phase out CFCs were entered into under the Montreal Protocol which was introduced in 1987. The most comprehensive initiative to date, however, was the 'Earth Summit' Conference on Environment and Development held in Rio de Janerio in 1992 (Johnson, 1993). With around 175 nations represented, with over 100 heads of state and government, and with over 1500 officially accredited non-governmental organisations, the summit is thought to have been the largest international gathering ever held. The conference debated a wide range of issues, grouped under the eight headings of atmosphere, biodiversity/biotechnology, institutions, legal instruments, finance, technology transfer, fresh-water resources and forests. Two international treaties, on climatic change and biodiversity, were opened for signature and were ratified by over 150 nations. A number of declarations and commitments to enter into further discussions were also made. In addition, a massive 600-page tome detailing a comprehensive agenda for the 21st century was adopted as a framework for future national and international steps in the fields of environment and development.

Agenda 21 comprises 40 chapters grouped into four sections. These cover (1) 'social and economic dimensions', (2) 'conservation and management of resources for development', (3) 'strengthening the role of major groups' (such as children, women and indigenous peoples) and (4) 'means of implementation'. Chapter 7 of section 1 has the most explicit urban focus as it outlines an agenda for promoting sustainable human settlement development. It incorporates sets of objectives for eight constituent programme areas (Table 22).

The overall human settlement objective of Agenda 21 of the Earth Summit is to improve the social, economic and environmental quality of human settlements and the living and working conditions of all people, in particular the urban and rural poor. It is envisaged that such improvement will be based upon technical cooperation activities, partnerships among the public, private and community sectors, and participation in the decision-making process by community groups and special interest groups. These approaches should form the core principles of national settlement strategies. In developing these strategies it is anticipated that countries will need to set priorities among the eight programme areas, taking fully into account their social and cultural capabilities.

The fact that an Earth Summit was held at all and that it led to several important agreements and declarations underlines the growing international concern for a sustainable future and a recognition that

Table 22 The Earth Summit: Agenda 21 Programmes for Promoting
Sustainable Human Settlement Development

Programme area: (A) Providing adequate shelter for all

Objective: the objective is to achieve adequate shelter for rapidly growing
populations and for the currently deprived urban and rural poor through
an enabling approach to shelter development and improvement that is
environmentally sound.

(B) Improving human settlement management

The objective is to ensure sustainable management of all urban settlements,
particularly in developing countries, in order to enhance their ability to
improve the living conditions of residents, especially the marginalised and
disenfranchised, thereby contributing to the achievement of national
economic development goals.

(C) Promoting sustainable land-use planning and management

The objective is to provide for the land requirements of human settlement
development through environmentally sound physical planning and land
use so as to ensure access to land to all households and where appropriate,
the encouragement of communally and collectively owned and managed
land. Particular attention should be paid to the needs of women and
indigenous people for economic and cultural reasons.

*(D) Promoting the integrated provision of environmental infrastructure: water,
sanitation, drainage and solid-waste management*

The objective is to ensure the provision of adequate environmental infrastructure facilities in all settlements by the year 2025. The achievement of
this objective would require that all developing countries incorporate in
their national strategies programmes to build the necessary financial and
human resource capacity aimed at ensuring better integration of infrastructure and environmental planning by the year 2000.

(E) Promoting sustainable energy and transport systems in human settlements

The objectives are to extend the provision of more energy-efficient technology and alternative/renewable energy for human settlements and to
reduce negative impacts of energy production and use on human health
and on the environment.

*(F) Promoting human settlement planning and management in disaster-prone
areas*

The objective is to enable all countries, especially those which are disaster-
prone, to mitigate the negative impact of natural and man-made disasters
on human settlements, national economies and the environment.

(G) Promoting sustainable construction industry activities

The objectives are, first, to adopt policies and technologies and to exchange
information on them in order to enable the construction sector to meet
human settlement goals, while avoiding harmful side-effects on human

health and on the biosphere, and, second, to enhance the employment-generation capacity of the construction sector. Governments should work in close collaboration with the private sector in achieving these objectives.

(H) Promoting human resource development and capacity-building for human settlements development

The objective is to improve human resource development and capacity-building in all countries by enhancing the personal and institutional capacity of all actors, particularly indigenous people and women, involved in human settlement development. In this regard, account should be taken of traditional cultural practices of indigenous people and their relationship to the environment.

Source: extracted from Johnson (1993: pp. 181–98).

coordinated global action is required. Critics have argued, however, that the summit achieved little because it failed to strike the necessary 'global bargain' between the developed and the developing worlds (Johnson, 1993). As envisaged, this bargain involves commitments from the developing countries over greenhouse gases, forests and sustainable development in return for concessions from the developed countries on finance, technology transfer and implementation. For Johnson (1993), the fact that greater agreement was not reached was because of the unwillingness of the developed countries to tackle their profligate lifestyles. This was coupled with the refusal of the developing countries to agree to limit the exploitation of their own natural resources so as to address the imbalance caused by the developed countries who have already used up so much of the earth's environmental capital and generated so much waste. The failure to reach agreements on a target date for stabilising emissions of greenhouse gases and on the protection of forests were the two principal failures of the conference.

It is far too early to identify and evaluate the practical outcomes of the Earth Summit. Such consequences will be slow to emerge and any effects will only be measurable in the long term. Particular difficulties surround Agenda 21, since its status is that of a framework for national action. No mandatory rules are specified and there are no clear targets against which progress can be measured. While it includes many practical suggestions to assist in achieving sustainable development, it does not have the power or resources to ensure implementation. The governments which attended the summit did not agree to transfer the necessary authority to any international institution. The onus of responsibility is upon individual nations, each of which is likely to respond differently. Few countries have systems of planning and management through which the necessary regulation can be introduced immediately, and many of those that do are committed to minimal intervention, in the belief that market mechanisms should decide. Many states have a tradition of acting independently and out of self-interest. The Earth Summit recognised sustainable urban development as an important goal for the next century, but further international agreement and commitment, and concentrated action, are required before rhetoric is likely to be translated into reality.

CONCLUSION

This book has attempted to analyse and account for the salient characteristics of the urban world and the global city. It has adopted a broad historical and geographical sweep, viewing urban development as a long-term process and seeing the world as being progressively, but as yet incompletely, interlinked and interconnected as an urban place in economic and social terms. Neither set of trends has run its course and though many urban changes have happened, especially in the last thirty years, more are in progress and are likely to occur. 'Urban world' and 'global city' are convenient catch-phrases which summarise dominant themes, but they describe conceptual ideals rather than the present situation.

The analysis and discussion in each of the eight chapters has identified key relationships and trends. The global urban population is concentrated in a small number of countries, especially China and India, although the percentage that lives in urban places is high throughout Latin America and the developed world. Within these regions large numbers live in million and mega-cities. Elsewhere, urban populations are smaller and towns and cities are fewer and more widely spaced. Urban places are the predominant form of settlement because they offer significant economies of scale, agglomeration and association. Their emergence reflects the power and persistence of processes which have concentrated large numbers of people in geographically small yet economically and socially viable communities. Wide variations, however, exist in the size pattern of cities in different countries from rank-size distributions at one extreme to primate at the other. Such differences probably reflect the strength of external linkages and, in the case of primate patterns, suggest that the integration of the global urban system is far from complete.

186

A hierarchy of urban places, with world cities at its head, extends across the globe, but separate and localised sub-systems which focus upon national primate cities are contained within it.

It is only comparatively recently that the concept of the urban world has begun to have any meaning. Urban development up to mid-century was largely restricted to the core countries of the world-economy. It was most advanced in those parts of north western Europe and North America which had been industrialised the longest and had dominated extensive political and economic empires. Elsewhere, urban development was embryonic, reflecting the widespread inability of pre-industrial economies to raise productivity and surpluses to levels necessary for significant and sustainable urban growth. When measured at the global scale the wholesale switch of population from rural to urban places is a phenomenon of the last thirty years. It is principally a product of changes in the distribution of population in developing countries. The factors involved are identified and accounted for by interdependency theory, which sees urban growth and urbanisation as consequences of the evolution of capitalism and its changing spatial relations. Of particular importance is the recent globalisation of the economy, a development which is reflected in and achieved through the rise of transnational corporations and global finance and producer services institutions, most of which are concentrated in a small number of world cities. The key development is the emergence of a new international division of labour in which production is dispersed to and so accelerates urban development in the peripheral areas within the world-economy.

As well as the shift into towns and cities, the world is progressively becoming urbanised in a social and behavioural sense. Traditionally, urban patterns of association and behaviour, though they themselves were highly varied, were a function of or related to place, being restricted to those who actually lived in cities. Today, the lifestyles and values of urbanites are being extended across the globe, both as a direct corollary of urban growth and urbanisation, and because they can be observed, copied and adopted in rural areas via telecommunications and the media. Urban images and messages, once largely Western in origin, are becoming more diverse as the producers of media products increase in number. The ability to participate in an urban way of life is becoming increasingly independent of location. The world is fast becoming a global urban society of which we are all residents.

The key issue which surrounds the urban world is whether it can continue in its present form. Urban patterns are now well established in most countries, but whether they can absorb a predicted doubling in the urban population over the next quarter century seems highly questionable. Doomsday scenarios have been invoked before and have come to nothing, but the sheer scale of likely growth suggests that they must be taken seriously this time. There is ample evidence that the global physical environment is being degraded and that many major cities are near to exhausting their abilities to cope with their exploding populations. Urgent and decisive action is required by governments if a sustainable urban future for all is to be secured.

This book has focused upon wide patterns and has addressed big issues. It has adopted a synoptic approach consistent with the aim of drawing together, overviewing and synthesising the literature, both established and recent, on world urban development. Some will question the wisdom of trying to generalise about the distribution, lifestyles and problems of half of the world's population. The global perspective and level of analysis, however, will be endorsed and accepted by those who recognise that a primary goal of the social sciences is to pursue and produce general understanding and explanation, though it will disappoint and frustrate those who revel in intricacies and fine detail. The overall purpose is to provide a broad framework within which local empirical work can be structured. The need for high-order generalisation is likely to increase as the pace of urban change quickens and as the pattern of global urban development becomes more complex. It took over eight millennia for half the world's population to become urban. Present predictions suggest that it will take less than eighty more years for this process to encompass the remainder.

RECOMMENDED READING

Breheny, M. J. (1992) *Sustainable Development and Urban Form*, London: Pion.

A useful collection of fourteen papers on all aspects of the urban sustainability debate. The principal focus is upon strategic planning for a sustainable future in the developed world. There are valuable case studies from the Netherlands, the United Kingdom and Sweden.

Blowers, A. (1993) *Planning for Sustainable Development*, London: Earthscan.

This book reports the views of leading planners on the policies and strategies which are required to ensure sustainable development. Despite the specific focus on the United Kingdom, the analysis and the practical suggestions have wider relevance especially to countries with well developed planning systems.

Cadman, D. and Payne, G. (1990) *The Living City: Towards a Sustainable Future*, London: Routledge.

A collection of thirteen papers which address issues of participation, ecological balance, local self-reliance, economic sustainability and technological choice.

Hardoy, J. E., Mitlin, D. and Satterthwaite, D. (1992) *Environmental Problems in Third World Cities*, London: Earthscan.

A comprehensive description and analysis of the environmental problems of cities in the Third World and how they affect human health, local ecosystems and global cycles. The authors further consider a range of practical solutions and how they could be implemented.

Haughton, G. and Hunter, C. (1994) *Sustainable Cities*, London: Regional Studies Association.

This book provides a comprehensive and detailed assessment of the concept of sustainability and how it applies to cities. Key themes are approached from geographical, ecological, economic and managerial perspectives. There is a valuable focus upon policy.

Johnson, S. P. (1993) *The Earth Summit: The United Nations Conference on Environment and Development*, London: Graham & Trotman.

This book is a compilation of the key documents associated with the Earth Summit, and the full text of conventions and programmes which were agreed. There is a useful preface by the editor in which he evaluates what was achieved and what further action is required.

Appendix

URBAN DEFINITIONS
AND DATA

Statistics on the urban world are assembled annually by the United Nations and are published in its *Demographic Yearbook*. They form the basis for the urbanisation tables which are included each year in the World Bank's *World Development Reports* and in numerous compilations of world facts and figures and geographical digests. The data are derived from national censuses; where these are unavailable or inadequate, they are based upon sample surveys or estimates. The characteristics of the data and the areas to which they refer are of critical importance in comparative urban analysis. Ideally, urban areas should be defined on the same basis and full censuses should be taken in each country at the same time on the same basis and with the same known levels of accuracy. The reality is, however, very different, with far-reaching implications for the validity of global urban study.

DEFINITIONS

Wide variations exist in the ways in which the populations of countries are divided into urban and rural. There is no standard approach because designation of areas as urban and rural is closely bound up with historical, political, cultural and administrative considerations. A wide range of criteria can be used, including: size of population, population density, distance between built-up areas, predominant type of economic activity, conformity to legal or administrative status, and characteristics such as specific services and facilities. Urban definitions tend to be revised infrequently and so they soon become outdated. This characteristic has important implications for studies of urban growth and urbanisation.

At first glance, population size would seem to be the most suitable indicator of urban status, but this measure is used in only twenty-six

190

of the 114 countries and sovereign territories in which census data are available (Table 23). Even with this simple criterion, wide variations exist as to the required minima. Places with as few as 200 inhabitants are considered to be urban in Iceland, while in Switzerland and Malaysia the lower limit is 10,000 people. Important differences exist between countries over whom to include in the urban population. Some censuses record all those who are present on a particular date, whereas others attempt to enumerate those who are normally resident. The latter results in grossly inflated figures where there are large numbers of people who are only nominally resident in the city but normally work and live away. This is the case in many developing countries where there are many migrant workers. It is especially serious in China where the city in which people are officially registered as living is often different to that in which they actually reside.

In twenty countries, population size is combined with other diagnostic criteria. Examples include Israel, Botswana and Zambia, which use population size and employment in non-agricultural occupations; Canada and France, which employ size and density criteria; and India, where consideration is given to population, density, legal and morphological characteristics. Such indices provide a more satisfactory basis for urban definition within the country concerned, although the composition and threshold values used again mean that the difficulty of making meaningful cross-national comparisons of urban patterns and problems is increased.

Urban places in the vast majority of countries are identified on a legal, administrative or governmental basis. No attempt is made to establish objective criteria. Instead, urban status is deemed to apply to all those places which occupy a particular level within the municipal hierarchy. This practice is especially common in small countries where towns and cities are designated by government decree. Thus, the urban places in Costa Rica are the administrative centres of cantons; in Pakistan they are places with a municipal corporation, town committee or cantonment; in South Africa they are places with some form of local authority; and in Iraq they are areas within the boundaries of municipality councils.

Global urban study is even less well served by fifteen of the world's national governments from which urban criteria in 1993 were 'not available'. The bases for urban definition, and how these compare with those in other countries, are therefore unknown. This group includes Colombia, Myanmar, Philippines, Somalia, Uganda and Belize.

Table 23 Definitions of urban places: selected examples

Criterion	Number	Examples
Population	26	Ethiopia: localities of 2,000 or more inhabitants. Venezuela: centres with a population of 1,000 or more inhabitants. Malaysia: gazetted areas with a population of 10,000 or more. Albania: towns and other industrial centres of more than 400 inhabitants. Iceland: localities of 200 or more inhabitants. Switzerland: communes of 10,000 or more inhabitants, including suburbs. Ireland: cities and towns, including suburbs, of 1,500 or more inhabitants.
Population plus additional criteria	20	Zambia: localities of 5,000 or more inhabitants, the majority of whom depend on non-agricultural activities. Botswana: Agglomerations of 5,000 or more inhabitants where 75 per cent of the economic activity is of the non-agricultural type. Canada: places of 1,000 or more inhabitants having a population density of 400 or more per sq. km. Israel: all settlements of more than 2,000 inhabitants, except those where at least one-third of households, participating in the civilian labour force, earn their living from agriculture. France: communes containing an agglomeration of more than 2,000 inhabitants living in contiguous houses with not more than 200 metres between houses, also communes of which the major portion of the population is part of a multicommunal agglomeration of this nature. India: towns (places with municipal corporation, municipal area committee, towns committee, notified area committee or cantonment board); also, all places having 5,000 or more inhabitants, a density of not less than 1,000 people per square mile or 390 per sq. km, pronounced urban characteristics and at least three-fourths of the adult male population employed in pursuits other than agriculture.
Legal, administrative governmental	54	Bangladesh: places having a municipality (*pourashava*), a town committee, *shahar* committee or a cantonment board. Pakistan: places with municipal corporation,

town committee or cantonment.
South Africa: places with some form of local authority.
Swaziland: localities proclaimed as urban.
Iraq: the area within the boundaries of Municipality Councils.
Mongolia: capital and district centres.
Brazil: urban and suburban zones of administrative centres of *municipos* and districts.
Uruguay: cities.

Not available	14	Myanmar
		Philippines
		Colombia
Total	114	

Source: United Nations (1992) *Demographic Yearbook*, pp. 150–3

DATA

As well as objectivity and consistency of definition, comparative international study is further impaired by variations in the frequency and timing of national population censuses. Following the lead established in Britain in 1801, most developed countries undertake a census of population, both urban and rural, every ten years. Censuses are held with due regard to the principles of social surveying and yield results with a known degree of statistical reliability. They form the basis for annual estimates of the size of the urban population. The reliability of such estimates is related to the time elapsed since the last census.

The practice elsewhere in the world is, however, more varied. In some countries, censuses are an occasional and infrequent rather than a regular occurrence. For example, the last census to be held in North Korea was in 1944! According to the United Nations, there are a further thirteen countries which have not held a national census since 1975 (Table 24). The list includes many of the poorest countries of Africa and Central America. Data on the present-day urban population of such countries are generated through extrapolation. Their reliability is especially low.

Variations in urban definition and in the availability and quality of urban data introduce significant but unknown biases into comparative urban study. They are of particular importance in analysing

Table 24 Major countries which have not held a
national population census since before 1975

Country	Most recent census
Africa	
Chad	1963
Djibouti	1961
Gabon	1961
Mauritius	1972
Namibia	1970
Nigeria	1963
Western Sahara	1970
Central America	
El Salvador	1971
Honduras	1974
Nicaragua	1971
Asia	
Cambodia	1962
North Korea	1944
Europe	
Iceland	1970
Netherlands	1971

Source: United Nations, 1994

contemporary urban change because the quality of definitions and data is commonly lowest in precisely those countries in which the urban population is largest and is growing the quickest. International standardisation of urban definitions and methods of data collection is a remote prospect. Global urban data are best regarded as crude estimates which support only general rather than precise statements about the distribution and growth of population in the contemporary urban world.

REFERENCES

Abu-Lughod, J. and Hay, R. (1979) *Third World Urbanization*, New York: Methuen.

Adams, R. M. (1966) *The Evolution of Urban Society*, London: Weidenfeld & Nicholson.

Akwule, R. (1992) *Global Telecommunications: Technology, Administration and Policies*, London: Focal Press.

Alcamo, J. and Lubkert, B. (1990) 'The city and the air: Europe', in Canfield, C. (ed.) *Ecocity Conference, 1990*, Berkeley CA: Urban Ecology, 12–20.

Alvarado, M. (1988) *Video World-Wide: An International Study*, London/Paris: UNESCO/John Libbey.

Alvarado, M. (1989) *Global Video*, London: UNESCO.

Barrios, L. (1988) 'Television, telenovelas and family life in Venezuela', in Lull, J. (ed.) *World Families Watch Television*, London: Sage, 49–79.

Becker, P. G., Frieden, J., Schatz, S. B. and Sklar, R. L. (1987) *Post-Imperialism. International Capital and Development in the Late Twentieth Century*, Boulder CO and London: Reinner.

Bekkers, W. (1987) 'The Dutch public broadcasting services in a multi-channel environment', in ESOMAR (eds) *The Application of Research to Broadcasting Decisions*, London: European Society of Market Research, 169–88.

Berg, L. van den, Drewett, R., Klassen, L. H., Rossi, A. and Vijverberg, C. H. T. (1982) *A Study of Growth and Decline*, London: Pergamon.

Berry, B. J. L. (1961) 'City size distribution and economic development', *Economic Development and Cultural Change* 9: 573–87.

Berry, B. J. L. (1971) 'City size and economic development', in Jakobson, L. and Prakash, V. (eds) *Urbanization and National Development*, Beverly Hills CA: Sage, 111–55.

Berry, B. J. L. (1973) *The Human Consequences of Urbanization*, New York: St Martin's Press.

Berry, B. J. L. (1976) *Urbanization and Counterurbanization*, London: Sage.

Berry, B. J. L. and Gillard, Q. (1976) *The Changing Shape of Metropolitan America: Commuting Patterns, Urban Fields and Decentralisation Processes, 1960–70*, Cambridge MA: Ballinger.

195

Berwanger, D. (1987) *Television in the Third World, New Technology and Social Change*, Bonn: Friedrich Ebert Stiftung.

Blowers, A. (1993) *Planning for a Sustainable Environment*, London: Earthscan.

Borchert, J. R. (1967) 'American metropolitan evolution', *Geographical Review* 57: 355–6.

Bowen, D. (1986) 'The class of 86', *Business* Nov: 34–41.

Bradnock, W. (1984) *Urbanisation in India*, London: Murray.

Breeze, G. (1966) *The City in Newly Developing Countries: Readings in Urbanism and Urbanisation*, London: Prentice Hall.

Brunn, S. D. and Williams, J. F. (1993) *Cities of the World: World Regional Urban Development*, New York: Harper Collins.

Cairncross, F. (1991) *Costing the Earth: the Challenge for Governments, the Opportunities for Business*, Boston MA: Harvard Business School Press.

Carter, H. and Lewis, C. R. (1991) *An Urban Geography of England and Wales in the Nineteenth Century*, London: Arnold.

Castells, M. (1977) *The Urban Question: A Marxist Approach*, London: Edward Arnold.

Castells, M. (1984) *The City and the Grassroots: A Cross-Cultural Perspective*, Berkeley CA: University of California Press.

Castells, M. (1989) *The Informational City: Information Technology, Economic Restructuring and the Urban-Regional Process*, Oxford: Blackwell.

Champion, A. G. (1989) *Counterurbanisation*, London: Arnold.

Chase-Dunn, C. (1985) 'The system of world cities: A.D. 800–1975', in Timberlake, M. (ed.) *Urbanization in the World Economy*, New York: Academic Press, 269–92.

Chase-Dunn, C. (1989) *Global Formation: Structures of the World-Economy*, London: Blackwell.

Childe, V. G. (1950) 'The urban revolution', *Town Planning Review* 21: 3–17.

Clark, D. (1974) 'Technology, diffusion and time-space convergence: the example of STD telephone', *Area* 6: 151–3.

Clark, D. (1979) 'The spatial impact of telecommunications', in Smith, R. C. (ed.) *Impacts of Telecommunications on Planning and Transport* Research Report No. 24, London: Departments of the Environment and Transport.

Clark, D. (1982) *Urban Geography: An Introductory Guide*, Beckenham: Croom Helm.

Clark, D. (1985) *Post-Industrial America*, London: Routledge.

Clark, D. (1989) *Urban Decline*, London: Routledge.

Cohen, R. B. (1981) 'The new international division of labour; multi-national corporations and the urban hierarchy', in Dear, M. and Scott, A. J. (eds) *Urbanization and Urban Planning in a Capitalist Society*, London: Methuen, 287–315.

Cross, D. F. W. (1990) *Counterurbanisation in England and Wales*, London: Avebury.

Cumming, S. D. (1990) 'Post-colonial urban residential change in Zimbabwe: a case study', in Potter, R. B. and Salau, A. T. (eds) *Cities and Development in the Third World*, London: Mansell, 32–50.

REFERENCES

Daniels, P. W. (1991) *Services and Metropolitan Development: International Perspectives*, London: Routledge.

Daniels, P. W. (1993) *Service Industries in the World Economy*, Oxford: Blackwell.

Davis, K. (1965) 'The urbanization of the human population', *Scientific American* 213: 40–53.

Davis, K. (1969) *World Urbanization*, Los Angeles CA: University of California.

De Stefano, J. S. (1990) 'Language, global telecommunications and values issues', in Lundstedt, S. B. (ed.) *Telecommunications, Values and the Public Interest*, Norwood NJ: Ablex, 52–72.

Dicken, P. (1992) *Global Shift*, London: Harper & Row.

Dissanayeke, W. (1988) 'Cultural identity and Asian cinema', in Dissanayake, W. (ed.) *Cinema and Cultural Identity*, Maryland: University Press of America.

Dogan, M. and Kasarda, J. D. (1989) *The Metropolis Era Vol. 1: A World of Great Cities*, London: Sage.

Dogan, M. and Kasarda, J. D. (1990) *The Metropolis Era Vol. 2: Mega-Cities*, London: Sage.

Drakakis-Smith, D. (1992) *Urban and Regional Change in Southern Africa*, London: Routledge.

Dwyer, D. J. (1979) *People and Housing in Third World Cities*, London: Longman.

El Shakhs, S. (1972) 'Development, primacy and systems of cities', *The Journal of Developing Areas* 7 (October): 11–36.

Elvin, M. and Skinner, G. W. (1974) *The Chinese City Between Two Worlds*, Stanford CA: Stanford University Press.

Feagin, J. R. and Smith, M. P. (1987) 'Cities and the new international division of labour: an overview', in Smith, M. P. and Feagin, J. R. (eds) *The Capitalist City*, Oxford: Blackwell, 1–37.

Fielding, A. J. (1982) 'Counterurbanization in Western Europe', *Progress in Planning* 17: 1–52.

Findley, S. E. (1993) 'The Third World city: development policy and issues', in Kasarda, J. D. and Parnell, A. M. (eds) *Third World Cities*, London: Sage, 1–33.

Fischer, C. S. (1976) *The Urban Experience*, New York: Harcourt Brace Jovanovich.

Forbes, D. and Thrift, N. J. (1987) 'International impacts on the urbanisation process in the Asian region: a review', in Fuchs, R. J., Jones, G. W. and Pernia, E. M. (eds) *Urbanization and Urban Policies in Pacific Asia*, Boulder CO: Westview.

Frank, A. G. (1967) *Capitalism and Underdevelopment in Latin America: Historical Studies of Chile and Brazil*, New York: Monthly Review Press.

Frank, A. G. (1969) *Latin America: Underdevelopment or Revolution*, New York: Monthly Review Press.

Friedman, Y. (1984) 'Towards a policy of urban survival' in di Castri, F., Baker, F. and Hadley, M. (eds) *Ecology in Practice Part II: The Social Response*, Dublin, Tycooly and Paris: UNESCO.

Friedmann, J. P. (1972) 'The spatial organisation of power in the development of urban systems', *Development and Change* 4: 12–50.

Friedmann, J. (1986) 'The world city hypothesis', *Development and Change* 17: 69–74.

Friedmann, J. and Wolff, G. (1982) 'World city formation: an agenda for research and action', *International Journal of Urban and Regional Research* 6: 309–42.

Frobel, F., Heinrichs, J. and Kreye, O. (1980) *The New International Division of Labour*, Cambridge: Cambridge University Press.

Fuchs, R. J., Jones, G. W. and Pernia, E. M. (1987) *Urbanisation and Urban Policies in Pacific Asia*, Boulder CO: Westview.

Fujita, K. (1991) 'A world city and flexible specialization: restructuring of the Tokyo metropolis', *International Journal of Urban and Regional Research* 15: 269–84.

Gans, H. J. (1962a) 'Urbanism and suburbanism as ways of life', in Rose, A. M. (ed.) *Human Behaviour and Social Processes*, London: Routledge & Kegan Paul, 23–38.

Gans, H. J. (1962b) *The Urban Villagers*, New York: Free Press.

Giddens, A. (1990) *The Consequences of Modernity*, Stanford CA: Stanford University Press.

Giddens, A. (1991) *Modernity and Self-Identity: Self and Society in the Late Modern Age*, Cambridge: Polity Press.

Gilbert, A. and Gugler, J. (1992) *Cities, Poverty and Development*, Oxford: Oxford University Press.

Giradet, H. (1990) 'The metabolism of cities', in Cadman, D. and Payne, G. (eds) *The Living City*, London: Routledge, 170–80.

Goddard, J. B. (1973) *Office Linkages and Location, Progress in Planning 1*, Oxford: Pergamon.

Goldfrank, W. L. (1979) *The World-System of Capitalism Past and Present*, Beverly Hills CA: Sage.

Goldstein, S. (1989) 'Levels of urbanisation in China', in Dogan, M. and Kasarda, J. D. (eds) *The Metropolis Era Vol. 1: A World of Great Cities*, London: Sage, 187–225.

Goldstein, S. (1993) 'The impact of temporary migration on urban places: Thailand and China as case studies', in Kasarda, J. D. and Parnell, A. M. (eds) *Third World Cities*, London: Sage, 199–219.

Goudie, A. (1989) *The Nature of the Environment*, Oxford: Blackwell.

Goudie, A. (1990) *The Human Impact on the Natural Environment*, Oxford: Blackwell.

Griffin, E. and Ford, L. (1993) 'Cities of Latin America', in Brunn, S. D. and Williams, J. F. (eds) *Cities of the World*, New York: Harper Collins, 225–66.

Gugler, J. (1988) *The Urbanisation of the Third World*, Oxford: Oxford University Press.

HABITAT (1987) *Global Report on Human Settlements*, Oxford: United Nations Centre for Urban Settlements and Oxford University Press.

Hahn, E. and Simonis, U. (1990) 'Ecological urban restructuring: method and action', *Environmental Management and Health* 2: 12–19.

Hall, P. (1966) (3rd edn 1984) *The World Cities*, London: Weidenfeld & Nicholson.

Hall, P. and Hay, D. (1978) *Growth Centres in the European Urban System*, London: Heinemann.

Hall, P., Gracey, H., Drewett, R. and Thomas, R. (1973) *The Containment of Urban England*, London: George Allen & Unwin.

Hardoy, J. E. and Satterthwaite, D. (1989) *Squatter Citizen*, London: Earthscan.

Hardoy, J. E. and Satterthwaite, D. (1990) 'Urban change in the Third World: are recent trends a useful pointer to the urban future?', in Cadman, D. and Payne, G. (eds), *The Living City*, London: Routledge, 75–110.

Hardoy, J. E., Mitlin, D. and Satterthwaite, D. (1992) *Environmental Problems in Third World Cities*, London: Earthscan.

Harvey, D. W. (1973) *Social Justice and the City*, London: Edward Arnold.

Haughton, G. and Hunter, C. (1994) *Sustainable Cities*, London: Regional Studies Association.

Hawley, A. (1981) *Urban Society: an Ecological Approach*, New York: Ronald.

Hay, R. (1977) 'Patterns of urbanisation and socio-economic development in the Third World: an overview', in Abu-Lughod, J. and Hay, R. (eds) *Third World Urbanization*, New York: Methuen.

Janelle, D. G. (1991) 'Global independence and its consequences', in Brunn, S. D. and Leinbach, T. R. (eds) *Collapsing Space and Time: Geographic Aspects of Communication and Information*, London: Harper Collins, 49–81.

Jefferson, M. (1939) 'The law of the primate city', *Geographical Review* 29 (April): 226–32.

Johnson, S. P. (1993) *The Earth Summit: The United Nations Conference on Environment and Development*, London: Graham & Trotman.

Johnston, R. J. (1980) *City and Society: An Outline for Urban Geography*, Harmondsworth: Penguin.

Kasarda, J. D. and Parnell, A. M. (1993) *Third World Cities*, London: Sage.

Kelly, P. M. and Karas, J. H. W. (1990) 'The greenhouse effect', *Capital and Class* 38: 17–27.

King, A. D. (1989) *Global Cities: Post-imperialism and the Internationalisation of London*, London: Routledge.

King, A. D. (1990) *Urbanism, Colonialism and the World-Economy. Cultural and Spatial Foundations of the World Urban System*, London: Routledge.

Kirkby, R. J. R. (1985) *Urbanization in China: Town and Country in a Developing Economy, 1949-2000 AD*, London: Croom Helm.

Knight, R. V. and Gappert, G. (1989) *Cities in a Global Society*, New York: Sage.

Knox, P. and Agnew, J. (1994) *The Geography of the World-Economy*, London: Arnold.

Lampard, E. E. (1955) 'The history of cities in economically advanced areas', *Economic Development and Cultural Change* 3: 81–102.

Lampard, E. E. (1965) 'Historical aspects of urbanisation', in Hauser, P. M. and Schnore, L. F. (eds) *The Study of Urbanisation*, London: Wiley, 519–54.

Langdale, J. V. (1991) 'Telecommunications and international transactions in information services', in Brunn, S. D. and Leinbach, T. R. (eds) *Collapsing Space and Time*, New York: Harper Collins, 193–214.

Levich, R. M. and Walter, I. (1989) 'The regulation of global financial markets', in Noyelle, T. (ed.) *New York's Financial Markets: The Challenge of Globalisation*, Boulder CO: Westview Press, 51–90.

Lewis, J. W. (1971) *The City in Communist China*, Stanford CA: Stanford University Press.

Leyshon, A., Thrift, N. J. and Daniels, P. W. (1987a) *The Urban and Regional Consequences of the Restructuring of World Financial Markets: The Case of the City of London*, Portsmouth: Working Papers on Producer Services no. 4, University of Bristol and Services Industries Research Centre, University of Portsmouth.

Leyshon, A., Thrift, N. J. and Daniels, P. W. (1987b) *Large Commercial Property Firms in the UK: The Operational Development and Spatial Development of General Practice Firms of Chartered Surveyors*, Portsmouth: Working Papers on Producer Services no. 5, University of Bristol and Services Industries Research Centre, University of Portsmouth.

Lull, J. (1995) *Media, Communication and Culture: A Global Approach*, Cambridge: Polity Press.

Lull, J. and Se-Wen, S. (1988) 'Agents of modernisation: television and urban Chinese families', in Lull, J. (ed.) *World Families Watch Television*, London: Sage, 193–236.

McGee, T. G. (1971) *The Urbanization Process in the Third World: Explorations in Search of a Theory*, London: Bell.

McGee, T. G. (1967) *The Southeast Asian City*, London: Bell.

McGlynn, G., Newman, P. and Kenworthy, J. (1991) 'Land use and transport: the missing link in urban consolidation', *Urban Futures* special issue 1, July: 8–18.

Machimura, T. (1992) 'The urban restructuring process in the 1980s: transforming Tokyo into a world city', *International Journal of Urban and Regional Research* 16: 114–29.

McLuhan, H. M. (1964) *Understanding Media: The Extensions of Man*, New York: McGraw-Hill.

McMichael, M. (1993) *Planetary Overload: Global Environmental Change and the Health of the Human Species*, Cambridge: Cambridge University Press.

Markusen, A. and Gwiasda, V. (1994) 'Multipolarity and the layering of functions in world cities: New York City's struggle to stay on top', *International Journal of Urban and Regional Research* 18: 167–93.

Mattelart, A. (1979) *Multinational Corporations and the Control of Culture*, London: Harvester Press.

Meadows, D. H., Meadows, D. L., Randers, J. and Behrens, W. W. (1972) *The Limits to Growth*, New York: University Books.

Meier, R. L. (1962) *A Communications Theory of Urban Growth*, Cambridge, MA: Massachusetts Institute of Technology Press.

Mollenkopf, J. E. (1993) *Urban Nodes in the Global System*, New York: Social Science Research Council.

Moss, M. L. (1987) 'Telecommunications, world cities and urban policy', *Urban Studies* 24; 534–46.

Moss, M. L. (1988) 'Telecommunications: shaping the future', in Sternleib, G. and Hughes, J. W. (eds) *America's New Market Geography: Nation, Region and Metropolis*, New Brunswick NJ: Centre for Urban Policy Research, Rutgers, 45–66.

Muller, P. O. (1981) *Contemporary Suburban America*, Englewood Cliffs NJ: Prentice Hall.

Noyelle, T. J. (1989) 'New York's competitiveness', in Noyelle, T. J. (ed.) *New York's Financial Markets: The Challenge of Globalisation*, Boulder CO: Westview Press, 51–91.

O'Connor, A. (1983) *The African City*, London: Hutchinson.

Odum, E. P. (1989) *Ecology and our Endangered Life Support Systems*, Sunderland MA: Sinaner Associates.

Organisation for Economic Cooperation and Development (1990) *Environmental Policies for Cities in the 1990s*, Paris: Organisation for Economic Cooperation and Development.

O'Riordan, T. (1989) 'The challenge for environmentalism', in Peet, R. and Thrift, N. (eds) *New Models in Geography: Volume One*, London: Arnold, 77–104.

O'Riordan, T. (1990) 'Global warming', *Marxism Today* July: 12–15.

Orrskog, L. and Snickars, F. (1992) 'On the sustainability of urban and regional structures', in Breheny, M. (ed.) *Sustainable Development and Urban Form*, London: Pion.

Pahl, R. E. (1965) *Urbs in Rure: The Metropolitan Fringe in Hertfordshire*, London: London School of Economics and Political Science Geographical Papers No. 2.

Pahl, R. E. (1968) *Readings in Urban Sociology*, Oxford: Pergamon.

Patterson, R. (1987) *International TV and Video Guide 1987*, London: Tantivy Press.

Pearce, D., Markandya, A. and Barbier, E. B. (1989) *Blueprint for a Green Economy*, London: Earthscan.

Perry, M. (1990) 'The internationalisation of advertising', *Geoforum* 22: 35–50.

Pool, I. de S. (1976) *The New Structure of International Communication: The Role of Research*, Leicester: International Association for Mass Communication Research.

Potter, R. B. (1985) *Urbanisation and Planning in the Third World*, Beckenham: Croom Helm.

Pred, A. R. (1977) *City Systems in Advanced Economies*, London: Hutchinson.

Preston, S. H. (1988) 'Urban growth in developing countries: a demographic reappraisal', in Gugler, J. (ed.) *The Urbanization of the Third World*, Oxford: Oxford University Press, 11–32.

Ramachandaran, R. (1993) *Urbanisation and Urban Systems in India*, Oxford: Oxford University Press.

Redfield, R. (1941) *Folk Culture of Yucatan*, Chicago IL: University of Chicago Press.

Reed, H. C. (1984) 'Appraising corporate investment policy: a financial center theory of foreign direct investment', in Kindleberger, C. P. and Audretsch, D. B. (eds) *The Multinational Corporation in the 1980s*, London: MIT Press, 219–43.

Rondinelli, D. (1989) 'Giant and secondary city growth in Africa', in Dogan, M. and Kasarda, J. D. (eds) *The Metropolis Era Vol. 1: A World of Great Cities*, London: Sage, 291–321.

Sassen, S. (1991) *The Global City: New York, London, Tokyo*, Princeton NJ: Princeton University Press.

Sassen, S. (1994) *Cities in a World Economy*, London: Pine Forge Press.

Schiller, H. (1976) *Communication and Cultural Domination*, White Plains: International Arts and Sciences Press.

Schteingrat, M. (1990) 'Mexico City', in Dogan, M. and Kasarda, J. D. (eds) *The Metropolis Era Vol. 2: Mega-Cities*, London: Sage, 268–329.

Sethuraman, S. U. (1981) *The Urban Informal Sector in Developing Countries*, Geneva: International Labour Organisation.

Sit, V. F. S. (1985) *Chinese Cities: The Growth of the Metropolis Since 1949*, Oxford: Oxford University Press.

Sit, V. F. S. (1993) 'Transnational capital flows, foreign investments and urban growth in developing countries', in Kasarda, J. D. and Parnell, A. M. (eds) *Third World Cities*, London: Sage, 180–98.

Skinner, G. W. (1977) *The City in Imperial China*, Stanford CA: Stanford University Press.

Smith, C. A. (1985a) 'Theories and measures of urban primacy: a critique', in Timberlake, M. (ed.) *Urbanization in the World-Economy*, London: Academic Press, 87–116.

Smith, C. A. (1985b) 'Class relations and urbanisation in Guatemala: towards an alternative theory of urban primacy', in Timberlake, M. (ed.) *Urbanization in the World-Economy*, London: Academic Press, 121–59.

Smith, M. P. and Feagin, J. R. (1987) *The Capitalist City*, Oxford: Blackwell.

Sreberny-Mohammadi, A. (1991) 'The global and the local in international communications', in Curran, J. and Gurevitch, M. (eds) *Mass Media and Society*, London: Arnold, 118–38.

Stevens, P. (1987) 'English as an international language', *English Teaching Forum* 25: 56–64.

Stoneman, C. (1979) 'Foreign capital and the reconstruction of Zimbabwe', *Review of African Political Economy* 11: 62–83.

Taaffe, E. J., Morrill, R. L. and Gould, P. R. (1963) 'Transport expansion in underdeveloped countries: a comparative analysis', *Geographical Review* 53: 503–29.

Taylor, P. J. (1993) *Political Geography: World-Economy, Nation-State and Locality*, London: Longman.

Thorngren, B. (1970) 'How do contact systems affect regional development?', *Environment and Planning* 2: 409–27.

Thrift, N. J. (1987) 'The fixers: the urban geography of international commercial capital', in Henderson, J. and Castells, M. (eds) *Global Restructuring and Territorial Development*, Newbury Park: Sage, 203–33.

Thrift, N. J. (1989) 'The geography of international economic disorder', in Johnston, R. J. and Taylor, P. J. (eds) *A World in Crisis*, Oxford: Blackwell, 16–79.

Timberlake, M. (ed.) (1985) *Urbanization in the World-Economy*, London: Academic Press.

Timberlake, M. (1987) 'World-system theory and the study of comparative urbanization', in Smith, M. P. and Feagin, J. R. (eds) *The Capitalist City*, Oxford: Blackwell, 37–64.

Tolba, M. K. and El-Kholy, O. A. (1992) *The World Environment, 1972–92: Two Decades of Challenge*, London: Chapman & Hall.

Torado, M. P. (1994) *Economic Development*, London: Longman.

Tracey, M. (1988) 'Popular culture and the economics of global television', *Intermedia* 16: 2–9.

Tracey, M. (1993) 'A taste of money: popular culture and the economics of global television', in Noam, E. M. and Millonzi, J. C. (eds) *The International Market in Film and Television Programs*, Norwood NJ: Ablex, 163–98.

Turner, R. (1962) *India's Urban Future*, Los Angeles CA: University of California Press.

Turner, R. K., Pearce, D. and Bateman, I. (1994) *Environmental Economics: An Elementary Introduction*, Hemel Hempstead: Harvester Wheatsheaf.

Ullman, E. L. and Dacey, M. F. (1962) 'The minimum requirements approach to the urban economic base', in Norborg, K. (ed.) *Proceedings of the I.G.U. Symposium on Urban Geography*, Lund: C.W.K. Gleerup, 485–518.

UNCTD (1991) *World Investment Report: The Triad in Foreign Direct Investment*, New York: United Nations.

UNCTD (1993) *World Investment Report, 1993: Transnational Corporations and Integrated International Production*, New York: United Nations.

UNESCO (1989) *World Communication Report*, Paris: UNESCO.

United Nations (1991) *World Urbanization Prospects*, New York: United Nations.

United Nations (1994) *Demographic Yearbook*, New York: United Nations.

Unwin, N. and Searle, G. (1991) 'Ecologically sustainable development and urban development', *Urban Futures* special issue 4, November: 1–12.

Vance, J. E. (1970) *The Merchant's World: The Geography of Wholesaling*, Englewood Cliffs NJ: Prentice Hall.

Vapnarsky, C. A. (1969) 'On rank-size distributions of cities: an ecological approach', *Economic Development and Cultural Change* 17: 584–95.

Varis, T. (1984) 'The international flow of television programs', *Journal of Communication* 24: 143–52.

Varis, T. (1993) 'Trends in the global pattern of television programs', in Noam, E. M. and Millonzi, J. C. (eds) *The International Market in Film and Television Programs*, Norwood NJ: Ablex, 1–13.

Wallerstein, I. (1979) *The Capitalist World-Economy*, New York: Cambridge University Press.

Wallerstein, I. (1980) *The Modern World-System 2: Mercantilism and the Consolidation of the European World-Economy, 1600-1750*, London: Academic Press.

Wallerstein, I. (1989) *The Modern World-System 3: the Second Era of the Great Expansion of the Capitalist World-Economy, 1730-1840s*, London: Academic Press.

Walters, P. B. (1985) 'Systems of cities and urban primacy: problems of definition and measurement', in Timberlake, M. (ed.) *Urbanization in the World-Economy*, London: Academic Press, 63–78.

Wardhaugh, R. (1987) *Languages in Competition, Dominance, Diversity and Decline*, New York: Basil Blackwell.

Warf, B. (1989) 'Telecommunications and the globalisation of financial services', *Professional Geographer* 41: 257–71.

WCED (1987) *Our Common Future*, Oxford: Oxford University Press.

Webber, M. M. (1964) 'The urban place and the non-place urban realm', in Webber, M. M., Dyckham, J. W., Foley, D. L., Guttenberg, A. Z., Wheaton, W. L. C. and Wurster, C. B. (eds) *Explorations into Urban Structure*, Philadelphia PA: University of Pennsylvania Press, 79–153.

Weber, A. F. (1899) *The Growth of Cities in the Nineteenth Century*, New York: Macmillan; 1962 reprint, Ithaca NY: Cornell University Press.

Wheatley, P. (1971) *The Pivot of the Four Quarters*, Chicago IL: The University of Chicago Press.

Wilhelmy, H. (1986) 'Urban Change in Argentina: Historical Roots and Modern Trends', in Conzen, M. P. (ed.) *World Patterns of Modern Urban Change*, Chicago IL: University of Chicago Research Paper 217–18, 273–92.

Williams, S. (1992) 'The coming of the groundscapers', in Budd, L. and Whimster, S. (eds) *Global Finance and Urban Living*, London: Routledge, 246–59.

Wirth, L. (1938) 'Urbanism as a way of life', *American Journal of Sociology* 44: 1–24.

World Bank (1992) *World Development Report, 1992*, New York: World Bank.

Young, M. and Willmott, P. (1957) *Family and Kinship in East London*, London: Routledge & Kegan Paul.

Zipf, G. K. (1949) *Human Behavior and the Principle of Least Effort*, New York: Addison & Wesley.

Zukin, S. (1992) 'The city as a landscape of power: London and New York as global financial capitals', in Budd, L. and Whimster, S. (eds) *Global Finance and Urban Living*, London: Routledge, 195–233.

INDEX

United Kingdom: counter-
urbanisation trends in 45, 52;
foreign direct investment by 84;
industrial capitalism in 68–9;
mass media in 123, 132;
monopoly capitalism in 73; rural
lifestyles in 111–12; suburban
lifestyles in 109; telecommunications in 158, 160; urban
development in 3, 18, 26, 45,
49, 62, 69; urban lifestyles in
108; world city in 141, 147,
150, 151, 152, 155, 159, 160
United Nations: media statistics
from 122, 128, 132; population
statistics from 14, 22, 46, 168,
171, 190, 193; sustainable
development and 182–6;
transnational corporation
statistics from 81–2
United States of America: counterurbanisation trends in 45, 50,
52; foreign direct investment by
84; mass media in 123, 125,
126, 128, 132; monopoly
capitalism in 74; suburban
lifestyles in 109–11; telecommunications in 158;
transnational corporations in
83; urban development in 3,
17, 26, 71; urban lifestyles in
104–6, 107–9; world city in
141, 147, 150, 151, 152
urban areas: definitions and
descriptive labels of 1, 21–2,
190–3; development of 1–2, 3,
5, 6; (counterurbanisation
trends 45, 50, 52, 56–7;
economic growth and development and 2–3, 21, 24, 28–9,
59, 60–74; environmental
problems and 172–86; future of
166–88; global 6–10, 59–74,
77–9, 88–97; socio-economic
aspects of 2–3, 5–6, 93–7;
stages of 52–7, 63–74, 77–9;
theories of 31–8, 57–63; trends
in 40–52, 55, 89–93, 166,
167–72) lifestyles in 2–3, 6,

100–6, 139; (mass media and
3, 102, 103, 121–34; studies of
101, 102, 106–14); linkages
between 27–8, 90; population
distribution in 1–2, 3, 5, 13,
14–25, 22–4; shanty towns
95–7; size distributions of 6,
25–31; social conflict in 24–5;
study of 4–6, 17–18; survey of
13–25; unreliability of data on
13–14, 193–4; world cities
137–42; (business information
and decision-making in 153–6;
case study of 161–3; command
and control functions in 138,
147–53; international finance
and 138, 142–7; telecommunications and 156–61); see also
suburbs
urban geography 4–6, 17–18
Uruguay 18

Vance, J. E. 65
Varis, T. 132
Venezuela 18, 123, 126–7, 133
Venice 59
Vietnam 20

Wallerstein, Immanuel 6
Walters, P. B. 27
Wardhaugh, R. 130
Warner Brothers 128
water supply 180–1
way of life see lifestyles
Webber, M. M. 119
Weber, Adna Ferrin 5, 69
welfare systems 24
Wheatley, P. 31, 36
Wilhelmy, H. 29
Wirth, Louis 104–6, 107, 112
World Bank 14, 49, 85, 190
world cities 137–42; business
information and decision-
making in 153–6; case study of
161–3; command and control
functions in 138, 147–53;
international finance and 138,
142–7; telecommunications and
156–61

Yemen 18, 43
Yucatán 106

Zagreb 175
Zaire 30, 125

Zambia 30, 191
Zimbabwe 24, 91–2
Zipf, G. K. 25
Zürich 150